To The Waters And The Wild

Petroleum Geology 1918 to 1941

By
Ellen Sue Blakey

Published By
The American Association of Petroleum Geologists
Tulsa, Oklahoma, U.S.A.
1991

(Frontispiece)
The Cities Service crew poses in the Seminole oil field. There were more horses than autos at that time.
Cities Service.

Copyright © 1988
Published 1991
The American Association of Petroleum Geologists
All Rights Reserved

Blakey, Ellen Sue
To the Waters and the Wild
    Bibliography: p.
      1. Petroleum—Geology—History   I. Title.
ISBN: 0-89181-804-9

Editors: Fred A. Dix, Robert H. Dott, Sr.,
        Ronald L. Hart, Anne H. Thomas
Production Manager: Ken Frakes
Special Projects Manager: Victor V. VanBeuren
Typographer: Eula Matheny

Contents

	Dedication	43	**Proper Tea and Water-Buffalo Milk**
	Preface	49	**Sojourn in South America (Robert Dott)**
1	**Introduction**	53	**Jungle Trails**
2	**The Big War**	82	**Working for the Survey**
4	**Bicycles Not Acceptable**	84	**Into a New Century**
8	**Move 'Em Out**	84	**Escape from Montana**
11	**Diamond Drills and Seismographs**	87	**The Pack Rat Syndicate**
12	**Fossils in the Earth**	88	**Too Much Ransom**
15	**A Geologist's Alibi**	90	**Blowout in Turner Valley**
19	**A Black Future**	91	**One Dead Horse**
20	**A Golden Land**	93	**Far Ends of the Earth**
20	**Signal on the Hill**	96	**Hacking a Trail Westward**
23	**Craters and Core Drills**	97	**Shooting the Rapids**
26	**Escape from the Firing Squad**	97	**The Big International Argument**
27	**Terse and Salty Language**	100	**Trouble with the Troops**
32	**Over the Top**	101	**No More Uncivilized Country**
34	**Dark Continent**	103	**Ask and Ye Shall Receive—Sometimes**
37	**Dangers of the Territory**	109	**Trouble with Allah**
40	**A Fiery Furnace**	109	**The Acid Test**

112	**Run Out of the Bean Field**	159	**Hospitality in the Andes**
112	**Riding Through an Oven**	164	**Two Hawks and a Greyhound**
119	**High Water and Barbed Arrows**	174	**Cognac on the Campfire**
121	**Comic-Opera War**	177	**Flour for the Camels**
123	**Flying High**	179	**Lost in a Sea of Sand**
129	**Hard Times in the Oil Patch**	181	**Flinging the Fantastic Toe**
132	**Goose Chase Blues**	185	**Wild Men of the Jungle**
134	**Scotch on the Rocks**	187	**Tiger in the Eye**
141	**Depression Down Under**	194	**Mountain High**
142	**Midst of a Revolution**	194	**Amid Falling Bricks**
143	**Coca for Pacha-Mama**	194	**A Matter of Luck**
146	**Mountain High**	195	**Turning Something Up**
146	**Confronting the Headhunters**	195	**Spirit of Adventure**
147	**On a Shoestring**	195	**The Important Thing**
148	**The Unsung Heroines**	197	**Too Long Too Deep**
149	**The Castor Oil Line**	197	**Fresh Enough for Scrambled**
150	**Reading by Turner Gas Light**	202	**End of an Era**
154	**The Academic Side**	206	**Acknowledgements**
157	**A Bad Frame of Mind**		

Dedication

When we began this work in 1983, we contacted as many of the old-guard petroleum geologists as we could locate, asking for photographs and stories. Some of them had long since retired and we had no records of them. It took a little digging, but once we found them, they overwhelmed us with their eager response.

When we explained that we only wanted materials prior to 1940, a number of people were upset: they felt that later material might well be lost—just as much of the earlier material had been—if something were not done about it soon.

They were right. We were thinking too small to begin with—and the material we received showed us that. For that reason, we decided to do a three-volume set instead of the one original volume planned.

One of those who responded was Ottmar F. Kotick who wrote us, "Many of the old guard are dead. Ergo it's high time for (a) history before the rest of us hunt new oilfields in wherever we are headed."

Kotick's remarks were appropriate. While we have been preparing the materials, at least one of the people who provided us with material went to hunt those new oilfields. We would like to dedicate this volume to those who have helped by providing their stories and photographs to be preserved for future generations—and to those who never had an opportunity to tell theirs. They have our gratitude—and may there always be an interesting horizon wherever they go, and an honest boulder for a pillow.

Preface

Most of the stories told here were supplied by geologists and their families. Some were from their letters, diaries and geological accounts. Others came from tapes and reminiscences. Still others came from works which they or others had published. Early newspaper accounts and company documents added still another dimension. But like all reminiscences, they are colored by personal points of view—by individual preferences and prejudices, by which side of the mountain or valley they happened to be standing on at the time. One geologist assures us that only his version is correct; another with an entirely different point of view vehemently responds that the first story could not be further from the truth. Those of us who only sift through words and listen cannot judge. We can only report. And if there are other versions we have not told, we have remained silent without malice but because we have not found those who told a different tale. Some very fine geologists left no personal accounts. Others' accounts are still buried with families we were never able to reach. So the stories we tell here may not be the most important geologically. Even though some may differ from what many believe to be factual, they certainly are indicative of the experiences of the early-day geologists in their quest for oil and gas. Our gratitude goes to those who responded to our request for materials. And our apologies go to the memory of others, those whose rich tales we never heard or found.

Field camp party from Ohio State University, July 1935, on top of Lookout Mountain overlooking Chattanooga, Tennessee. Left to right: Robert Lockett, Daniel A. Busch, Gordon Pringle, Willard Phelps, Joseph Kelley, Dick Priddy (?), John Frye, Roland Snow, Jack Sparks and Robert F. Eberle. Photo taken by Paris Stockdale, the director of the field camp.

To The Waters And The Wild

To The Waters And The Wild

Wars are marker stones along the path of history. They are significant because they signal change. The daily cloth of living is woven with the threads of new technology and in renewed concern for life and its relationships.

Before the world was engulfed in war in 1914, the petroleum industry was still in its infancy even though it had attracted plenty of attention. There was something hypnotic about oil. The showy gushers and blowouts were always good for a crowd, if not for the surrounding environment or the oilman. But there was often a glut on the market simply because the supply in many areas exceeded the demand. Certainly the potential was there.

The average laborer was just beginning to be able to afford the automobile; the tin lizzie was still a plaything of the wealthier urban class. Mechanized farm machinery was a dream come true, sure to outdistance the plodding horse. But it had not yet made inroads into the great Midwestern heartland where we grew so much of our foodstuffs.

(Preceding page)
By 1920, geologists were to be found everywhere in Oklahoma. Shorty Shaffer (left) and Kelly Van Dyne, from Ohio State University, share a commodious running board in Walters Field, Cotton County. E.W. Owen/Mirva C. Owen.

Airplanes were fascinating to watch as cocky pilots barnstormed and demonstrated loop-the-loops at remote air fairs. But there was as yet no regular air passenger or mail service in the country. Most of the factories still thrived on coal, and the thick black smoke belching from the stacks was a sign of civic prosperity.

When Germany declared war, the need for petroleum heated up as armies took to the skies and navies to the seas. Petroleum literally fueled the fight. America sailed to victory upon a sea of oil, newsmen said.

But there was a second, long-term effect. In the factories geared up for wartime production, men developed new technology. When the war ended, there was a different, peacetime economy, and this economy knew and understood the expanding technology and was ready to put it to use. Oil and gas were no longer geological stepchildren, but the darlings of that new industrial technology.

Between World War I and World War II, petroleum geology lured young men to the ends of the earth. They chopped their way through unexplored tropical rainforests and hung by pickaxes from cliffs. They crossed deserts on camels and dug 20-foot holes by hand for samples and fossils. Some suffered physical breakdowns. Some were killed by unfriendly tribesmen; others lost their lives in the roaring rapids of jungle rivers. A few were caught in petty battles and larger wars, cut off from civilization for months at a time.

It was not easy to endure such hardships. It took no small man to make it through those days of geology by foot. It required men almost larger than life to find the underground riches, men of endurance and drive, men tall in dreams and long in dedication, men to match the mountains.

The Big War

World War I erupted suddenly on July 28, 1914. Although the United States was not at first directly involved, it was an important supplier to the Western Allies.

The economy was stimulated by wartime needs—particularly the energy industries—which were called upon to fuel the war. Geologists were especially in demand to provide military training. Fredrik Thwaites, at the University of Wisconsin, was one of several geologists who taught plane-table mapping to army trainees during the war.

Joseph Theophilus Singewald, Jr., and B.L. Miller, professors at Johns Hopkins University, explored South America in search of minerals for industrial wartime use. Their trip was cut short when war conditions in Europe began to affect South American politics.

Some geologists of military age were given special dispensation to remain in the states in search of oil and gas to fuel the war.

The Mid-Continent area—particularly Kansas, Oklahoma and Texas—drew special attention. The exploration efforts of Empire Gas and Fuel Co. had paid off tremendously at

Augusta and El Dorado, Kansas, where giant fields had been discovered in 1914 and 1915, respectively. It was a major victory for geologists. Empire executives were so impressed that they ordered their geological department immediately to increase its staff before the competition gathered up all the available geologists.

Just south of the Kansas border and the El Dorado field was Oklahoma's Osage Indian Reservation. Prior to 1916, oil leases on the reservation were let by sealed bids. That year, the bidding process was turned over to a skilled and colorful auctioneer known as Colonel Walters. The oil companies were eager for those leases, and it became customary for top officials of many oil companies to attend the sales. Sometimes the bidding turned into a free-for-all as rivalries heated up. Geologic guidance was often left behind.

There had to be some irony—and some sense of justice—in those sales. The Osage Indians, whose ancestors had been given what had then appeared to be the worst possible land, sat stoically, watching as the white man's millions rolled into the tribal coffers.

The world war drew closer to home in 1918 when J. Elmer Thomas, newly elected president of the American Association of Petroleum Geologists, enlisted in the Army Signal Corps' Air Service. Soon after he was commissioned major in the Aerial Photographic Section, he visited Bartlesville, Oklahoma, and persuaded several of the geologists working for Empire Gas and Fuel Company that the Aerial Photo Section offered the best military-service opportunity for geologists. Some followed his example and enlisted, including Walter R. Berger, Robert H. Dott, Earle P. Hindes, Charles C. Hoffman and John M. Nisbet.

The Secretary of the Interior decreed that the young geologists were more important in the discovery of oil than they were marching off to war. Several were able to participate in the U.S. Geological Survey project to map the geology in Osage County.

The men worked in teams—a geologist and an instrument man. Groups of several such teams worked out of centrally located camps. Both William Argabrite and Kenneth Heald were active in the Osage area. Clarence Ross remembered field work in wintertime. "Certain problems were presented by field work during winter, under war-time conditions and pressures," he wrote. "Snows at times stopped work and the roads were impassable at other times. Our 'tin lizzies' did not always run and were commonly reinforced (and held together) by 'bailing wire.'

"Mapping was done on a planetable, using a telescopic alidade and stadia 'rod.' Position or location of the planetable, as well as the elevation of many points, were determined by rod 'shots,' and, under some conditions, by triangulation."

"Precise planetable work is time-consuming," wrote Kenneth Heald, "therefore, procedures that would hasten the work of the instrument men were devised." This included use of a good set of lungs. "It was generally held that Frank Clark could be clearly heard and completely understood at a distance of a half mile," wrote Heald.

"The use of the mirror on the Brunton compass was particularly helpful (for

Shell Oil geologist F.E. Vaughan poses in shirt-sleeves in August, 1918, near Controller Bay, Alaska. The Bering glacier and Chugach Range are in the extreme distance.   L.M. Clark/AAPG.

communication) in heavily wooded areas, where branches and even trees had to be cleared away to free the vision of the instrument man. The geologist would follow a rock ledge through the wooded area to a point he wanted put on the map. He would climb a tree and attempt to locate the instrument man with his field glasses, and then flash his mirror to attract the instrument man's attention. The instrument man would read the distance from the stadia rod and determine the elevation of the geologist's feet in the tree, the geologist would measure and record the distance of his feet above the ledge so that between them the correct elevation of the point could be determined for the final map.

The team of Frank Clark and Paul V. Roundy made a fine record in both area covered and accuracy in their Osage County work. Most of the time, they used a buckboard drawn by a team of big horses. One of those horses was not very sure-footed, and would occasionally stumble and fall, then jump to its feet, running as it hit all fours.

"Clark, with his background of life on a frontier farm, naturally was the driver of the team," Heald wrote. "One night he came in with his hand badly gashed and bleeding from several deep wounds...The stumbling horse had fallen and lunged to his feet, jerking the buckboard so suddenly that Clark was thrown out. However, he kept hold of the reins with one hand, and with the other hand reached out and grabbed a barbed-wire fence and did not let go. He explained it very simply: 'If I had not held those reins, the horses would really have run away and there was no telling what would have happened to Roundy!' "

Bicycles Not Acceptable

In Texas, geologists were working overtime, as an unprecedented number of wells were drilled on their recommendations. Rio Bravo Oil had used geologists since 1902, and consistently continued to do so. In 1916, it was the only corporation whose drilling and exploration were both under the supervision of geologists.

Humble Oil increased its geological department to three men in 1918, and then to ten, carrying on an aggressive exploration and development program in north Texas under Wallace Pratt.

The Ranger, Texas, field was crowded with geologists, drillers, speculators and assorted camp followers. Inadequate living quarters and lack of roads gave it the nickname "Ragtown," the "rag" meaning canvas (tent) town.

The main road from Strawn to Ranger was a dusty trail clogged with horse-drawn wagons loaded with oil-field equipment. Bandits periodically held up travelers at the wire gates that separated ranches, then moved on to town to shoot up the gambling joints and honky tonks. It was no place for prissy manners nor academic vocabularies.

One geologist who worked the area was James J. McGowan, a gigantic Irishman who had been trained as a London policeman before taking up geology. He became something of a

In August, 1918, Shell Oil was already exploring Alaska. F.E. Vaughan snapped Leslie M. Clark, "gun bearer etc.," as he posed near Controller Bay. The Bering glacier is in the extreme distance.
L.M. Clark/AAPG.

The ruts that passed for roads in Texas in 1919 were tough on automobiles. One geologist swore the most important man on the survey trips was the mechanic. Without him, the geologist was forced to repair his own axles and differential gears. K.C. Heald/U.S. Geological Survey.

local celebrity. He disarmed one notorious outlaw and was known to whip teamsters who would not yield half the road to his model-T Ford.

Those who preferred more refined surroundings looked to Fort Worth. Although it was far from the oil fields and served by poor roads, it became headquarters for most geologists and operators. J. Elmer Thomas had his elegant office in the Fort Worth Club. Consultants from other regions were frequent visitors in the city.

Men such as Tom Harrison and his young associate, Arthur Eaton, came down from Denver by train in a stateroom fitted up as an office. They would spend the day in town and then turn north to Tulsa in the evening.

**Nitroglycerine was dangerous to handle and transport, and the men who worked with it had nerves of steel. But the government needed oil to fuel the war, and "nitro" was just part of the activity in the Osage County area.
Cities Service Co.**

Despite the growing number of professionals, there were still plenty of people who passed themselves off as geological experts but who had no scientific background. Many managed to find oil.

In 1918, Alexander Deussen, an independent geologist, noted, "Very few of the wild-cat wells in the Coast country are located upon the advice of really competent geologists. We have a great many pseudo-geologists, wiggle-stick men and others of one kind or another who locate many of these wells. Many of them are located on the basis of no sound indication. Even in the case of one of the larger companies, locations are made without respect to valid indications. Most of the larger companies, however, are guided by sound geological advice in their drilling operations."

Some international companies had trouble providing sound advice of any kind—simply because they did not understand conditions in the United States. They commonly made the mistake of trying to run the American operation from European headquarters, such as London or The Hague. Few understood the problems of space and distance in the wide-open country.

In one cost-cutting measure, Roxana Oil (Royal Dutch/Shell) ordered all oil scouts to turn in their autos and use bicycles. Pete Fears was scouting out of Duncan, Oklahoma, at the time. Some of his wells were separated by 75 miles of sandy or muddy roads. He resigned the same day he received the order.

A similar problem with foreign management occurred in California. Ben van der Linden had been detailed by Royal Dutch Petroleum to look over the oil situation in California for their subsidiary, the American Gasoline Company. They decided to hire a geologist, and J.E. "Brick" Elliott was recommended. He may have been the first American geologist employed full time by the Royal Dutch/Shell Group in the United States. Elliott's first assignment was to accumulate all available oil data in California. In the spring, he and van der Linden made a motor reconnaissance trip from Coalinga to San Diego and back up through the San Joaquin Valley. Although they carefully studied the Santa Maria oil field, both seemed to favor the Los Angeles area. Elliott's final report was sent to The Hague. Over the next few years, Royal Dutch spent nearly $3 million in the Ventura field without any results. In 1918, The Hague dispatched Wilhelm van Holst Pellekaan to California to sabotage or liquidate the Ventura field operation. In an effort to liquidate the property and sell to some unsuspecting company, Pellekaan suggested that Elliott change his original report favoring Ventura. Elliott refused. Pellekaan was determined to get them out of the area. He then offered the leased area east of the Ventura River to Associated Oil Company. Cort Decius, Associated's chief geologist, promptly accepted. Van Holst then attempted to give away the remainder of the leased area west of the Ventura River. At that point, van der Linden stepped back in to prevent further giveaways. It was fortunate for Royal Dutch/Shell that he did, for the field in its entirety produced more than two billion barrels of oil after it was brought in.

The El Dorado, Kansas, field was one of the first fields to be discovered with extensive use of geology. Haulers and owners line up—including one World War I doughboy—as mules and horses head out for the fields.   Cities Service Co.

Move 'Em Out

World War I officially ended in 1918. It demonstrated the strategic importance of oil supplies, altered political alignments and changed the economic posture of almost every nation. When Russia was isolated by the 1917 Russian Revolution, the United States found itself the most powerful military, economic and moral force in the world. "Fresh American troops and materiel, shipped en masse over a submarine-infested ocean, had supplied the margin of victory," Ed Owen wrote. "Airplanes, tanks and mechanized transport had finally overwhelmed the German foot-soldiers after four years of stupid trench warfare. Oil had become the prime strategic material, the indispensable agent of mobility, the most tractable source of energy."

Its use extended beyond military power. Domestic consumption was up. All oil wells were

producing wide open. But the new fields—Ranger, Breckenridge and Burkburnett—were not keeping up with the decline of the older giants—Cushing, Healdton and El Dorado. America began to look abroad. Success in Mexico led geologists into Central and South American countries, particularly to Panama, Costa Rica, Guatemala and Honduras.

Others struck out for such wild areas as the East Indies. They were not always successful. Jersey Standard spent $6 million in the Indies, only to average a miserable 135 barrels per day in 1918.

In March 1919, the American Association of Petroleum Geologists held its fourth annual meeting in Dallas. The subject was the urgent need for efficient oil exploration. David White, chief geologist of the U.S. Geological Survey, gave the keynote address. "I wish to

encourage, I wish to quicken the oil and gas geologists who are here, to the discovery of new fields," he said. "I would have them urge their companies to take a deeper and broader interest in the acquisition of oil reserves in other parts of the world, and 'do it now.' The American oil companies would do well to take a larger leaf from the book of foreign companies, and assiduously follow the example of the latter in acquiring oil reserves in any part of the earth where oil is likely to be discovered, and that is everywhere; for there is no part of the globe, this side of the poles, from which oil will not be brought into use if it can be found there."

Suddenly geologists were in demand. Not only were there new areas to explore, but a mountain of geological appraisal work had been created by a federal income-tax law, which provided that properties whose value was increased by oil discoveries could be re-evaluated as to oil reserves, as a basis for depletion allowances.

To complicate matters, the practice of petroleum geology, itself, had grown more sophisticated—beyond the ken of amateurs—as geologists searched for such hidden features as buried mountains, arches and uplifts beneath the surface. "So desperate were the oil companies for employees with experience in science," said Lewis Weeks in his autobiography, *A Lifelong Love Affair*, "that there was a joke that soda jerks were being hired to go to the oil fields to show where to drill. Being that close to the pharmacy department, it was said, had given them some 'scientific training.' "

Diamond Drills and Seismographs

Numerous advances occurred in the industry, some as the result of technology developed for the war, others as applications from one industry filtered to another.

Marion M. Travis, president of Midco Petroleum Company, became interested in core-drilling methods after seeing them in the Colorado mines. Core drilling offered an opportunity to locate and determine elevations on key marker beds in covered areas, and could serve as an extension of surface mapping.

In 1918, Travis had helped design and manufacture a mobile diamond drilling rig operated by a gas engine. It was used in 1919 to drill in the Chilocco Indian Reservation, Kay County, Oklahoma, and the Billings field.

As a result of this success, the prestige of geologists in Oklahoma went up. New companies began to crop up, including Amerada Petroleum Corporation, formed in the United States by Britain's Lord Cowdray.

In 1920, at the annual AAPG meeting, J.A. Udden proposed "a slight adaptation of some of the present seismographs" which might send a seismic wave "started by an explosion at the surface of the earth" and which might record the "emerged reflection of this wave" on an instrument "placed at some distance from the point of explosion."

Nearly a dozen geologists from Empire Gas and Fuel Co. joined the war effort. J. Elmer Thomas (first president of AAPG) recruited them into the Aerial Photo Section of the Signal Corps. Among them were these seven stalwarts seated on the steps of the field house, Cornell University, Ithaca, New York, in August 1918. Left to right: John Millar, Robert H. Dott, Earle P. Hindes, John M. Nisbet, Charles C. Hoffman, Lloyd H. Pasewalk and Raymond Leibensperger.

(Opposite)
Nitroglycerine was loaded into the "torpedo," which the remote-control-operated cable then lowered down the well hole to the depth of the producing formation. In Oklahoma, where this well was drilled, that depth rarely exceeded 2000 feet.   Cities Service Co.

It was a clear indication of what was about to happen in the industry. Those who had expected the demand for petroleum to go down after the war were more than surprised. Almost 34,000 wells were drilled in 1920; production and distribution facilities became strained to the limit. Standard Oil Company (Indiana) had 945 filling stations which were out of gasoline an average of 12 days, each, during the first eight months of 1920. In Pennsylvania, crude topped out at $6.10 although it was a mere $2.15 in California and $3.50 in the Mid-Continent. The U.S. Geological Survey kept peering through the gloom to announce that domestic production would soon be in an "inevitable decline." A financial panic throughout the nation did not help matters. Still, all but two of the largest oil companies had geological departments by this time. And the fields were producing.

In May, the giant Burbank field was discovered in Osage County, Oklahoma. Southern Arkansas had gotten into the act when a well was drilled that flowed 40 million cubic feet of gas. Oil seemed everywhere.

If geologists were finding more oil, it was because they knew more about it. Aerial surveys and aerial photographs were the latest tools for working surface geology. During the war, planes had been used for reconnaissance photos and surveys. In 1919, a war-time airplane was used to fly over California's Richfield and Santa Fe Springs areas. The photographer used a hand-held camera as the plane swung over the areas. Fairchild Aerial Surveys was the pioneer whose planes set the standard for the industry. "In blazing desert the aerial camera finds its evidence often clearly exposed," Fairchild wrote. "In tropics, obscured by jungle mantle, the camera lens still reveals the topography and shows the inquisitive geologist much. So much, in fact, that he has created a new classification in his ranks, the photo geologist. Where the going is easy, aerial photography helps him. Where a region is inaccessible or impossible to explore thoroughly from the ground, the geologist finds his aerial photography indispensable."

Fossils in the Earth

Another new tool in the geologist's kit was subsurface geology and the study of micropaleontology—the new buzzword of the industry. E.T. Dumble, chief consulting geologist of Southern Pacific Railroad (Pacific Oil Company, Associated Oil Company), had become aware of the occurrence of identifiable specimens of Foraminifera in well cuttings, and of their usefulness for correlation in the oil fields of Mexico. He hired Esther Richards (Mrs. Paul Applin) who had done some work for Southern Pacific in California, and put her in charge of the first industrial micropaleontology laboratory for the Rio Bravo Oil Company at Houston in 1920. The Marland Oil Company of Mexico employed Joseph A. Cushman, a leader in such research, during 1922-1924. J.J. Galloway was another eminent micropaleontologist. Micropaleontology was particularly useful on

In 1918, K.D. White was assigned to Bogotá, capital of Colombia, South America. There he encountered heavy rains, slow trains and even slower helpers. K.D. White/American Heritage Center, University of Wyoming.

the Gulf Coast. Humble Oil's staff, under Wallace Pratt with Alva Ellisor, was particularly strong in paleonteology as was The Texas Company with Hedwig Knicker.

Subsurface studies of stratigraphy, paleogeography and geologic history became critical factors in oil exploration, particularly in the Mid-Continent region. Geologists were able to understand and explain more about the occurrence of petroleum.

To A.I. Levorsen, a paleogeologic map portrayed a buried, unconformable surface, just as an areal geologic map portrays the modern surface. He coined apt phrases to make paleogeologic maps come alive—"flash-backs" into geologic history, "phantom faces" of Mother Earth, faces that persisted only a moment in geologic time, faces pieced together from "isolated glimpses" of different segments commonly widely separated from each other.

These maps gave him what he liked to call "a worm's eye view."

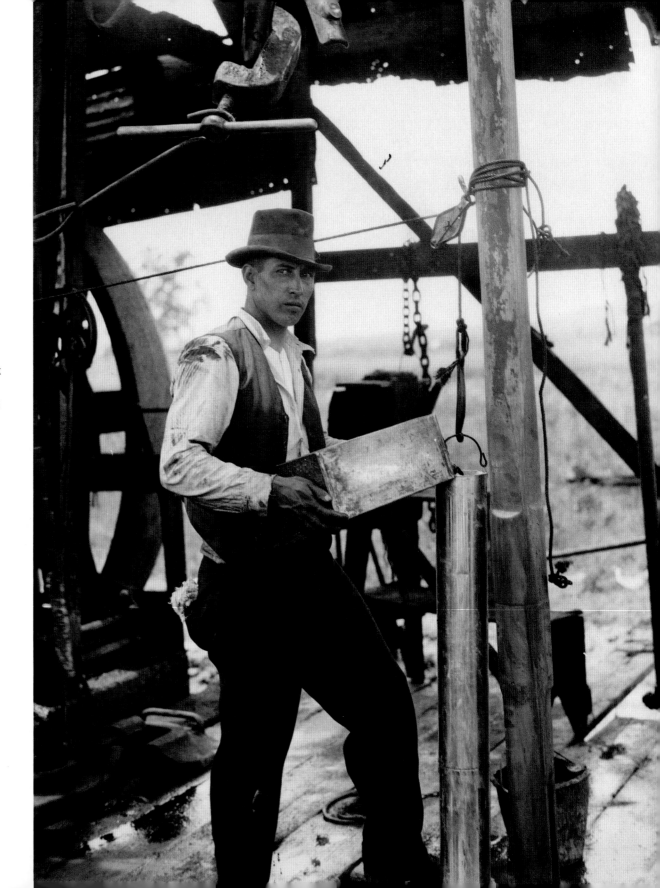

"Well-shooting" with nitroglycerin was commonly practiced to enhance production. The explosion sent a network of fractures through the producing formation, and the fractures facilitated oil flow through the sandstone or limestone to the drill hole.   Cities Service Co.

Wallace Pratt took another major step when he hired a young lady as a summer intern. There had been other women who studied geology. Charles Gould's geology classes at the University of Oklahoma had always had a few of the fairer sex, as did classes in many colleges and universities. But now, here was a woman actually working *in the field*. This was notable for two reasons—first, there were not even many men employed as geologists at the time. Second, it was against all tradition—in fact, it was almost heresy, considering that for years drillers had insisted it was a sure sign of trouble when a woman set foot on a rig floor. Now here was a slip of a thing actually in the field.

Alva Christine Ellisor had begun work examining well cuttings in the Ranger field, about 1919. She was particularly adept at micropaleontology and its application to oil drilling. She worked diligently to improve records of drillers' logs—still inaccurate and unreliable, at best. Fragments of megafossils which were recovered were generally unsatisfactory. Her work with Foraminifera and their significance on the Gulf Coast provided one more method by which geologists could track underground oil patterns.

Not everyone believed the method worked, however. In December 1921, Esther Richards [Applin] presented a paper at the Paleontological Society, in Amherst, demonstrating the use of Foraminifera to establish the age of several subsurface formations around the salt domes of south Texas. Professor J.J. Galloway rose to contradict her. "Gentlemen," he said, "here is this chit of a girl, right out of college, telling us that we can use Foraminifera to determine the age of a formation. Gentlemen, you know that it can't be done!" A year later, Galloway advertised his own ability to do what he had said could not be done.

Kelly Van Dyne poses in the Walters field, Cotton County, Oklahoma.
E.W. Owen/Mirva C. Owen.

A Geologist's Alibi

All but one or two of the large companies turned to geologists to help them find oil. For the next decade, geologists opened numerous fields by identifying structures—anticlines, domes and faults. Although there were still fields that were opened up without the help of—and some in spite of—geologists, it was no longer reasonably easy to find oil. The easy discoveries had already been made.

Down in Oklahoma, geologists like Sidney Powers were scouring the countryside looking for drilling sites. "I do not favor the western country west of a line between Lawton and Frederick and will continue geological work east of this line unless instructed otherwise," he wrote Everette DeGolyer in May, 1920. In July he wrote, "I wish to call your attention to the great possibilities of production in Love County...I suggest the blocking of a large tract of land somewhere west or northwest of Cheek for the purpose of drilling at least three tests to a depth of at least 2000 feet and probably deeper. My plan would be to collect, very carefully, samples from these wells, in order to determine the location of hidden folds parallel to the Hewitt-Healdton line. If evidence of such a fold was found another large oil field could be discovered. If no evidence was found the investigation would be purely a gamble. One of the other companies is planning a drilling campaign of this kind and certain Ardmore people are intending to drill in this sand desert as soon as their holdings at Hewitt are drilled up...Acreage is still to be procured at Hewitt by very careful buying direct from the farmers."

Wallace Pratt was less enthusiastic about Mid-Continent production than Powers. "What propaganda!" Pratt wrote Powers in July. "The truth is that Oklahoma-Kansas production has slumped, and Oklahoma will have to take a new spurt to make 300,000 bbls/daily. Texas is producing more oil right now than Oklahoma! One well in Texas is making as much oil as the largest field in Oklahoma!! The Breckenridge field is making twice as much oil as any Oklahoma field!! Except for Hewitt (which is Texas to HO&RCo.) what has Oklahoma to offer except a few erratic Mississippi limestone wells? Nowhere in which a decent pool (not even Gypsy of Boston) has developed.

"And now in Kansas with all the wells that have gone dry through the Boone Chert you hear the stupendous news to seize on the new Chautauqua County wonder to bolster up your sinking case. I can take nine wells like some we have in Texas and make more oil than the whole state of Oklahoma makes.

"And surface geology works no better in Oklahoma with the general practitioners lauding it than does surface geology in Texas...And several domes east of Tulsa are spiked with dry holes! North central Texas will make an average of 100,000 barrels daily for four years! What Oklahoma field ever did or ever will do that? Or Louisiana field either? All from one sand and one structure. Rapid decline? Rapid decline?? How about the toboggan at Duncan?

"Anyway, I notice that Oklahoma geologists when in search [of] geology come to Texas for it."

Despite Pratt's thumbs-down attitude toward Oklahoma, he congratulated Powers on Amerada's new production. "Who but you could have had nothing one week and 1200 bbls the next week?" he wrote. "Who else ever got 1200 bbls in one year with one million? Tell Mrs. Powers what I think of you."

Even when geologists did spot what they believed to be legitimate locations, they were not always praised. Willis Storm was in Carter County's Hewitt field in the fall to help develop Wirt Franklin's properties. "In most of the dry holes around the Hewitt field, most of which were drilled with cable tools, the cuttings were pretty good," Storm noted. "I finally decided that the trap structure was both folding and faulting...There was considerable disagreement with my analysis by the more learned geologists that, by inference, indicated that faulting was a lazy geologist's alibi. Nevertheless, Swigart and Schwartzenbek confirmed my reasoning in their Bureau of Mines report on the Hewitt Field."

H.L. Doherty, founder of Cities Service, visited Bartlesville in 1919 to promote the war effort. He was an honored guest of the Cities Service Engineering Research Department, including the women employees of this pioneering research group.   Cities Service Co.

Oil men gather under the Million Dollar Elm near Pawhuska, Oklahoma, during a lease auction in the early 1920s.
Phillips Petroleum.

A Black Future

Oil fields were still going strong in Texas after the war. In 1920, geologist Julius Fohs was working with Colonel A.E. Humphreys at Mexia, Texas, where a local gas company had already drilled a well. Fohs induced Humphreys to drill deeper, and struck oil. Wallace Pratt had actually ignored the Mexia district, and as late as April 1921, he reported to Humble, "We do not consider this well or this District as particularly significant...We are watching it and expect to take care of the situation if we feel any investments are warranted." When it became apparent that Mexia was a major find, Pratt was chagrinned. "My employers should have fired me," he wrote, "but they did not. They did ask me what we should do. My reply was that we should go to work to learn something about Mexia, which we had failed to do earlier."

Pratt wrote a friend, "The future was black and my stock was at a low ebb." He sent one of his recruits, Dwight Edson, a geologist fresh out of Dartmouth, to study the terrain. He came up with some data that were so promising, yet so confusing in the light of accepted geological doctrine, "...that I traveled up from Houston and joined him at his home in Mexia. We immediately set to work in his kitchen late Saturday night. Mrs. Edson had gone to bed. We were tired and frustrated, but still puzzling over our astounding data when suddenly the light broke and there was the complete explanation, simple and plain as day."

What Pratt and Edson confirmed was that the really rich oil deposits in the area were not trapped by the anticline, but by geological faulting. Prior to then, many geologists had generally assumed that faulting was not a means of containing oil but of allowing it to escape from a formation.

Pratt was so excited that, although it was two in the morning, he telephoned William Farish, head of Humble, and asked permission to lease as much land as possible. "We had no competition," Pratt wrote. "No other company realized the facts for weeks to come. The last 100-acre tract and one most favorably situated belonged to Dr. Thompson of San Antonio. He was canny enough to realize his strategic advantage and to demand a bonus of $2,500 per acre plus a 1/6 royalty—exhorbitant [sic] terms. At first, I talked to him at San Antonio over the telephone. Later, as negotiations dragged out, I think he came to Mexia. He would not come down [on his price] and I could not persuade Humble to risk a quarter-million dollar gamble." Pratt finally convinced Humble to take 50 percent in the lease and Gulf to take the other 50 percent. The gamble paid off. Out of 180 wells drilled on Pratt's leases, 175 produced oil. "The Thompson tract...produced more than 1,000,000 barrels of oil...Mexia made Humble geology and geologists respectable again."

A Golden Land

California seemed a remote region, apart from the rest of the continent. Most California geologists were graduates of either Stanford University or the University of California, Berkeley; they were a tight-knit group, and a majority held both engineering and geology degrees. When outside companies tried to break into the market, they most often found their entry much smoother if they hired men who were already familiar with local conditions and personalities.

Academic geologists found California particularly intriguing. Its geology and the concept about it was constantly changing as the earth shifted along California's fault lines. The man who became best known for following California's quirks was Bailey Willis, who had joined the staff at Stanford University in 1915. Willis had been chief geologist for the U.S. Geological Survey for several years prior to joining the university. Most people knew him as the earthquake expert. In those days, ideas such as earthquake weather and seismological cycles were still in vogue. Willis went contrary to popular opinion on many geologic theories. He insisted that the world was heating up, not cooling off, and would break through its crust some time in the future—in about two billion years. In the fall of 1919, Willis lectured his structural geology class on oil production, consumption and reserves, and wound up by telling them, "The world is at your feet."

It seemed that way. Gasoline rationing was concrete evidence that there was indeed a scarcity of oil. The demand for new reserves sounded as if it meant an increased demand for petroleum geologists—and attractive starting salaries. The reaction among the students was immediate. Enrollment in the field geology course totaled 61—an all-time high. The shortage did not, however, translate into those dreamed-of riches. When Maxwell Van Clief graduated, he was unable to get a job and finally settled for one as roughneck on a drilling crew. Arthur F. Turman was lucky—he started as a petroleum geologist in 1920 and felt fortunate to receive $150 a month.

Signal on the Hill

J.E. Elliott had recommended drilling the Long Beach-Signal Hill area several years before and had been turned down by Royal Dutch-Shell, for whom he worked at the time. He and Wilhelm van Holst Pellekaan, who had been put in charge of the geological department, had hardly seen eye to eye. Another young geologist, Frank Hayes, also recommended the area; but Pellekaan was prompt in condemning it. Four holes had been drilled on the nearby Dominguez Hills structure without favorable results. Hayes pointed out that Signal Hill was a much greater structure and that it seemed safe to assume that in the earlier stages of the structural development of both, most of the oil had migrated to Signal Hill.

Pellekaan would have none of it and insisted that Hayes drop the issue. Pellekaan took most of the men out of California to work elsewhere and left Alvin Theodore Schwennesen, a Stanford graduate, to carry on exploration. Pellekaan began work in Mexico. He did not expect much to happen in California. Schwennesen was just there "to keep a geologist on hand just in case anything happened." Schwennesen busied himself looking over what had been done; and in the spring of 1920, he ran across Hayes's report on Signal Hill. He employed Dwight Thornburg to check over Hayes's map and make a report. When he saw Thornburg's favorable report, Schwennesen went to check for himself.

There were some expensive homes on Signal Hill at the time, and it was developing into an exclusive residential area. Leasing was difficult, but Schwennesen did it. A well was sunk—Shell's Alamitos No. 1. Not everyone was convinced it was a producing area. W.W. Orcutt had drilled a dry hole for Union on the hill and felt certain there was nothing there. When Shell started drilling, Orcutt remarked to a friend, "I'll drink all the oil there is at Long Beach!" To his chagrin, the Alamitos No. 1 came in about June 1921, opening the greatest field in the basin. No one held Orcutt to his bet.

Pellekaan was suddenly back in California, with a much changed attitude, and the activity in California took on a new slant. Wells went down everywhere. Signal Hill was alive with drillers, pipeliners, wooden rig builders and all the usual personalities and paraphernalia that accompany an oil boom.

"Nick the Greek ran the Spud-In Cafe on Obispo Street near Bob and Butch McKeon's welding shop," Garth L. Young recalled. "The cafe was a tent stretched over a wood frame with counter and kitchen all in one. It ran 24 hours a day. Nick never seemed to sleep. He was always ready with short orders or anything else that was wanted. Cocoanut cream pie was his specialty. About twice a week, sometimes in the daytime, sometimes at night, the place was robbed. The robbery was so common that no one paid much attention as to the method or procedure. Nick, however, got tired of the raids after a while and decided to do something about them. He made a deal with the city of Signal Hill police to put in a red light on a pole a block from the cafe, with a switch in his place so that he could alert the police when a robber had just left his place with the receipts. They caught every one of the robbers within a few weeks. That ended the robbery problem, and Nick thereafter enjoyed a life of peace and security."

Single men lived in bunkhouses or boarding rooms, "so they had nothing to do in their spare time but go searching a cure for their singleness," noted one old-timer. "Rooms were scarce so I holed up with a few other roughnecks in the garage on Cherry Avenue on the south slope of the Hill," Garth Young remembered. "We all boarded in the same place which served meals at all hours. Not long after my rent started, Union's well hit a big gas pocket just behind the garage. Pipe blew out of the hole and wrapped around everything nearby, including my garage quarters. No one was hurt, but mud and water from the well fell on the roof, leaked in and almost collapsed the structure. The well blew wild for several days, but fortunately did not catch fire. However, enough mud cascaded down the Hill to cover the Pacific Electric train tracks four feet deep, suspending train service for a week or two. As suddenly as it blew out, the well sanded up again and flow stopped. Then we all moved back into the garage and resumed sleeping there a few hours each night."

Company families fared well. Company houses were $20 a month with gas, water and light free. Gardeners were hired to keep the families in flowers and vegetables and allow time for golf, bridge or a camp movie show. Several of the wells were drilled by George F. Getty, J. Paul Getty's father. "This old man was a typical oil man with leather puttees, jacket, cap and steel-blue eyes," Young said. "How he stood the strain of constant work is still a mystery to those who knew him."

Young's boss was a man by the name of Sam Mosher. "He also was the man we all depended on to keep us supplied with money for expansion," he said. "Sam had many friends who became interested in buying stocks in oil ventures. He would show up anytime day or night with an interested prospect and, using a stop-watch, he would measure the production rate into a five-gallon can filling it from the 'Look Box' near the storage tanks. He would calculate the 24-hour production from the number of minutes it took to fill the can. It was an impressive way to raise money, and he never failed. Sam was a super salesman."

Financing the projects was often at least as much trouble as finding them. E.L. Doheny, well known for his work in California, was like all the rest of the oil men. Doheny was said to work 24 hours a day and was "always looking ahead." But according to the California old-timers, he had his ups and downs, most of them monetary. One old-timer named Hunt recalled an occasion when Doheny found himself "temporarily embarrassed." Hunt went down to Doheny's home on Beaudry Street to get his pay; but when he arrived, he found several men standing around on the front porch, smoking cigars and talking to each other. They were obviously creditors. Hunt quietly made his way to the back and up the stairs. He knocked on the door, called softly for Doheny and identified himself. Doheny came to the back door and let him in. "Have you got 15 or 20 cents for carfare?" Doheny asked. Hunt responded that he did, and Doheny laid out his plan. "I want you to stomp around and make noise inside the house and keep those buzzards on the front porch occupied. I'm going to sneak out the back door here and go downtown and raise some capital and see if we can get this thing going again." There was no use asking for his pay, so Hunt lent Doheny the carfare and thus ended up with less money when he left than he had when he arrived. But Doheny soon rounded up more capital and was back in the hustle again.

Craters and Core Drills

Prior to the war, California had produced more oil than Oklahoma or Texas. But even in some of the best areas, it was hard to convince companies to drill. When Ralph Arnold had taken the first lease in the Santa Fe Springs area, he could get no major oil company interested. Union Oil, guided by geologist William W. Orcutt, had been trying for a decade—first with cable, then with rotary tools—but without any luck. Union had actually been trying for 12 years to make it big; and in 1922, they struck the Alexander gas well. The blowout swallowed the derrick and scattered drill pipe for a quarter of a mile. "A workman who was on top of the derrick got as far down as the double board when the top was carried away. He was picked up in the sump sixty feet away but with only minor bruises. Large rocks thrown out by gas kept laborers hundreds of feet away for days, and the gas kept enlarging the crater by eating into its sides. A crater formed even around the shallow water well of an orange grove two hundred feet away and swallowed a nearby derrick and storage tank. A week after the blowout, fine gray sand carried up by the gas was over everything within a radius of a mile, and for a quarter of a mile in every direction the earth in spots seemed porous and perforated by gas seeps."

The activity and the special conditions in California brought about development of core drilling for rotary wells. J.E. "Brick" Elliott, a seasoned geologist, noticed that the majority of the wells in the area were drilled by rotary rigs, and operators had no way of locating beds from which great amounts of salt water were infiltrating into the oil sands. He developed a double-cylinder core barrel; and in 1922, he formed his own company.

Cities Service employees—Doherty Men, Doherty Auxiliary, Doherty Girls—stage a war effort parade in Bartlesville. The DMF was the largest industrial fraternity in the United States at the time.
Cities Service Co.

The Seminole field helped establish Oklahoma as the country's leading oil producing state in the 1920s. But rains turned the roads to a heavy gumbo mud that became an obstacle course to man, beast and model T.   M.D. Maravich.

Elliott provided his own core-drillers. The crew were top hands and were paid tool-pusher salaries for an eight-hour day, plus a ten-dollar bonus for a successful core. On call 24 hours a day, their average take ran approximately $1,000 a month—a plum salary in those days.

Elliott also offered another service that set a precedent in the fields. At that time, there were still few wells being drilled under continuous observation by geologists or engineers. Elliott offered to spend a week or ten days constantly observing the drilling of any well that was scheduled to be cored. The service made more companies and their drillers aware of the value a well-site geologist.

Escape from the Firing Squad

Lewis Weeks returned to Mexico after World War I, where he was put in charge of all geology and engineering in the Southern Mines Division of the Greene Cananea. Conditions had changed little from his first time in the country. Bandits, weak government and competition combined to make a bad situation. "Periodically, the bandits would descend on the banks and the Chinese shops and relieve them of their money and portable goods," he wrote in *A Lifelong Love Affair*. "They also took the mining company's horses and shot the Chief of Police. The Mayor escaped from the bandits in his nightshirt. You never knew who among the mining employees was friend and who was foe. I was dismayed to be told that one of the Mexican workmen with whom I was beginning to learn to speak passable Spanish was one of them."

When conditions grew worse, General Diaz, the military governor of the state, was called to Cananea to take charge. But it was too late. "In a skirmish with the bandits, the government forces came off poorly," Weeks wrote. "A few of the bandit leaders were arrested trying to escape over the border and were brought back and tried. Those convicted were summarily hanged in a nearby grove of trees. One who had been held particularly responsible was left to dangle from a metal pipe over the main street of the town." Weeks decided the trials and tribulation were too much and returned to the states.

K.D. White was undaunted by such activity. He had moved north from geological work in South America, into Tampico. He completed a hurried reconnaissance of the oil fields from Tepetate south to Cerro Azul. "On Friday, May 16, while on my way to Cerro Azul accompanied by Mr. Baron and the truck driver I was stopped by a detatchment [sic] of Villistas or rebels, on the Cawdray [Cowdray?] Oil Co. pipe-line road about three miles north of Cerro Azul, and robbed of a saddle, bridle, blanket, two canteens and about $20.00 in cash," White wrote Dr. B.W. Dudley in New York. "The soldiers offered us no indignities, contenting themselves with taking everything in sight, though they were kind enough to leave me my note-book and instruments. We were expecting to spend the night at Cerro Azul and were prepared for a two day stay at that camp. The soldiers or bandits—the two being synonymous—refused to permit us to pass so we were forced to return to Los Naranjos." The band of rebels had been active around Cerro Azul for several weeks. Good guides refused to work in the area, knowing the chances of being robbed and the probability of the mules being taken. So White postponed the trip. "For the present we will visit those districts that are unmolested," he wrote.

Because of the unsettled conditions, many oil companies cabled their geologists to return to the states for safety. One geologist headed back for the Texas border on a dinky little mountain railroad which connected to the Mexican National Railway. When revolutionaries stopped the train, the naive geologist protested. He was thrown off with a half dozen others while his baggage was carefully investigated. Like most geologists, he carried a gun—a "take-down model 22-caliber rifle" and several hundred rounds of ammunition. To the revolutionaries that meant he was a spy. He was lined up with other, less savory characters, his hands tied behind his back, and marched up against the side of a corrugated iron shed. A firing squad formed, and a corporal proceeded down the row asking each man if he wished to be blindfolded. When he reached the American, he looked twice, then turned to his leader. Here was a mistake, the Mexican corporal asserted. The young man was not a spy—he was a geologist for whom the corporal had worked in Tampico. The American stared, and then recognized the peon who had had charge of a mule train for a surveying party he had taken into the jungle a few years before. They had become close friends during the trek.

The corporal finally convinced the bandit leader to spare the young man's life. He was allowed to go back to the train, sans money and baggage. As the train pulled out, he watched through the window as the other six men were shot, one by one.

Geologists were not the only ones who were expelled from the country. William Buckley, former lawyer and independent oil operator in Tampico, headed the Pantepec Oil Company of Mexico. He was accused several times of plotting to overthrow the Mexican Government, which did not see eye-to-eye with his interests. He was expelled from Mexico in 1921 as a result of his political maneuverings.

Terse and Salty Language

South America appeared to be a land of petroleum promise, and oil companies began sending in geologists almost as soon as the war was over. William Argabrite began work for Standard-Vacuum Oil Company (Standard Oil of New Jersey and Socony), in Venezuela and Colombia. In December 1919, Kenneth C. Heald and Kirtley F. Mather crossed the Eastern Andes from Cochabamba to Santa Cruz, Bolivia. Heald and Mather had a pack train of mules; and they studied the oil seepages in the vicinity of Santa Cruz and southward.

It was called "getting a horse," and skeptics still thought the horse would win out. So did some geologists, especially along the Neuquen River bluffs, Argentina. R.H. Dott.

Before the war, Joseph Theophilus Singewald Jr. and B.L. Miller, both professors at Johns Hopkins University, had visited all the leading mines in South America. The war had stopped their exploration, but as a result of their work, they had published *Mineral Deposits of South America* in 1919. It earned them overnight professional prestige, and they were back in the country almost as soon as conditions permitted.

Singewald sailed with Professor Edward W. Berry, a Johns Hopkins paleontologist, on a geological research tour of South America. They rode and hiked from the crest of the Andes to the nitrate plains of Chile. In northern Peru, Singewald carried out an oil examination for Shell Oil Company.

"Singewald never exalted his adventures or strayed far from geological facts or theories in discussing his trips," according to J. Brian Eby. "His colleague, Berry, however, was given to terse and salty language." Berry remembered an unforgettable lunch with primitive Andean Indians. "The meal consisted of one large pot of what appeared to be meat soup," he said. "After downing some of the concoction, Singewald, inadvertently and belatedly, stirred the contents of the pot with a stick and lifted to the surface the body of one boiled dog." Singewald spoke impeccable Spanish, but Berry prided himself on knowing neither Spanish nor Indian dialects. Instead, he favored vernacular English. The only time Singewald ever saw the Andean Indians jump to the job was when Berry directed a few well-chosen, old-fashioned American swear words in their direction. Berry explained that he had used the only international language he knew.

Harvey Bassler poses in the Andes in 1921 shortly after he began work in Peru. His later work was almost entirely in the Amazon Basin. American Museum of Natural History.

A young Huasteca boy poses beside the statue of a saint. "All big churches in the Huasteca had an idol with head broken off to show Indians that their gods could not help themselves," Price wrote.
W.A. Price.

In 1920–21, W. Armstrong Price led a party into Veracruz, Mexico, for Transcontinental Petroleum (Standard of New Jersey). His assistant, "Slats" Gibson, was handy with a camera, and they photographed the natives. One group at Tantoyuca performed a "furious dance" they called a "Juapango." "One Indian stripped to a breech clout and did a really furious shimmy with tiger tail," Price wrote.   W.A. Price.

(Left)
The hostess's 15-year-old daughter, Innocente, danced a fandango at afternoon tea on the patio.   W.A. Price.

W. Armstrong Price and "Slats" Gibson are ready for the field with leather belt and suspenders, notebook, Brunton compass, handlevel, mini-Kodak, aneroid, hammer and—last but not least—a .45 handgun. Gibson was a former lieutenant and had served at the Château Thierry, France, in World War I.   W.A. Price/AAPG.

(Left)
A column of government troops marched into Tantoyuca in 1920–21 as Standard Oil (New Jersey) members looked on. The troops came to execute rebel General Tomas Isquierdo, who broke his parole. Some of the geologists witnessed the execution. "The general got a cigarette, was shot, tumbled into an open grave [and was] covered up," Price noted. "A preview execution (?) left him for dead on [the] ground, but a woman salvaged him," he added.   W.A. Price.

Over the Top

Geologists had been eyeing northern Canada for several years. They knew there were oil seeps 1,000 miles north of the Alberta foothills, near a remote trading post known as Norman Wells.

Dr. T.O. Bosworth, an English geologist, had examined the properties and reported that the seepages were not very impressive, although the Fort Creek shales and Beavertail limestone were remarkable. But there were some adverse conditions that would have to be overcome if anyone were to be successful in finding oil. It was said that in summer, mosquitoes were big enough to shoot down with a rifle. In winter, temperatures sank to 60° below zero. That posed problems for both men and machinery.

Imperial Oil Limited, a subsidiary of Standard Oil of New Jersey, had set up an exploration and production department in 1914, with Bosworth as chief geologist. But Imperial did not get around to exploring the Norman Wells area until July 1920.

The group was under the direction of 23-year-old Theodore (Ted) Link, who had joined the staff in 1919. His entourage consisted of eight men and an ox, plus a cable-tool drilling rig with ancillary supplies and equipment. Their mission was to set up drilling operations and conduct geological studies.

They set out from Edmonton by rail travelling 300 miles to the Peace River. Once they reached the Peace River, their trials really began. It was 1,600 miles down winding rivers to Norman Wells—down the Peace River by scow to the Slave River at Lake Athabasca, down the Slave to Fort Resolution on Great Slave Lake, 100 miles across the lake, and a thousand miles down the Mackenzie River, a cold, clear stream that lumbered along at only five miles an hour but ranged four to ten miles wide. It included a four-mile portage on the Peace River and a sixteen-mile portage that took fifteen days at Smith Rapids on the Slave River. They journeyed for two months, and it was the first time that such massive equipment had traveled down the Mackenzie.

The Canadian North was harsh country, and equipment had to be transported long distances which included rapids and long portage paths. Anything that could accomplish two uses was welcomed. Theodore A. Link painted one canoe to serve as a stadia rod while working along the MacKenzie River, near the discovery well at Norman Wells, N.W.T., Canada in 1919. T.A. Link/AAPG.

In early September they put their ox, Old Nig, to work hauling wood from the steep banks of the river to build a cabin, storehouses and a stable, and to stack wood for the coming winter. Then there was the derrick, boiler and engine to get up the hill. The job was nearly complete when the river steamer made her final trip to pick up those who were returning to civilization for the winter. Five of the party, including Link, returned with the steamer. Three remained at Norman Wells to act as watchmen and to get the hole started before the bottom dropped out of the thermometer.

When the men drilled to about 300 feet, they found oil and gas shows. But the long winter night quickly set in, and for 10 months, the party remained isolated in their log cabin, a building which the *Imperial Oil Review* described as "an exclamation point in 2,000 miles of frozen silence."

The river was sheathed in a 10-foot coat of ice. By Christmas, the ox had to be served as stew. From that point on, supplies grew thinner. It was May before Link and seven others could set out for Norman Wells. This time, they had with them 20 tons of equipment in two scows and a motor boat.

The expanded entourage did not make the trip easier. One scow was lost in the rapids on the Peace River. The entire party almost went under in the Smith Rapids on the Slave River. The motorboat went aground on sand bars several times, and had to be freed by men working in icy, waist-deep water. By the time they reached Fort Resolution, the one remaining scow had been stripped to an essential 16 tons of equipment. It was 100% overloaded and leaking badly, and there was still 100 miles of open water on Great Slave Lake before they even reached the Mackenzie.

It was July 8, 1920, when they arrived at Norman Wells—none too soon for the men at the campsite, who had been reduced to living on fish and flour since June. According to Link, they were "almost crazy."

One of the young soon-to-be geologists who joined Link on the geological reconnaissance trips along the MacKenzie and its tributaries was Alex McQueen. "Geological work was confined to a study of the formations along the river and tributaries," Link wrote later. "Plane-table...traverses were made, and not without difficulty. Mosquitoes and black flies, although not as bad as last year, made life miserable, nevertheless. Inland trips had to be made with bedding, grub, planetable, alidade, tripod and rod on our backs. Tents were eliminated as excess, and many times we got a good soaking from the rain. Too much food is also a hindrance to inland trips. Hardtack and bacon grease are the staples."

Link mapped an anticline and A.Q.P. Patrick set out to drill on its crest. While drilling proceeded, Link continued his investigations. In August, he set out for a tributary creek on the Mackenzie River. "On August 23, 1920," reported the *Imperial Oil Review*, "when the drill was at a depth of seven hundred feet, word was brought to Mr. Link that oil was standing in the casing...within a few feet of the surface. 'Don't bother me again until it over-flows,' said the geologist, busy at his work." Four days later, a breathless Patrick again

Theodore Link shows off for the camera in 1918 while working in the Cisco, Texas, area. AAPG.

sought out Link. The well was at 783 feet, and oil was flowing over the top. Link returned to the drill site, and for 40 minutes they watched as a fountain of oil shot 70 feet into the air. Then satisfied, they shut it in and capped the well.

There was no radio or air service out of Norman Wells. The closest telegraph service was Fort MacMurray, Alberta, which was six weeks of up-stream boat travel from the site. The group set out, barely ahead of another winter. They crossed Lake Athabasca the night it froze over for the season. "We had to buck ice floes all the way up the Athabasca River to Ft. McMurray, and finally walk the last 30 miles, when the ice floes became too thick," they reported.

It was not until they reached the fort in October that the outside world heard the news. Norman Wells was a producer. It was the first production from the prolific Devonian reef reservoirs, the northernmost commercial oil production in the Western Hemisphere and the first significant discovery in Western Canada.

The work did not get any easier as they continued to survey the area, and improvisation was a way of life. In order to complete the first trip to the site by air, in 1921, the men had to fashion an airplane propeller from a sled runner and moose-hide glue.

Dark Continent

As companies stretched out in search of oil, they eyed Africa with great curiosity. Perhaps, as with Mexico and South America, her riches lay just beneath the surface. When the war in Europe ended, travel and government red tape smoothed out. Oil men once more struck out for new territory. In September 1919, inveterate adventurer Kessack Duke White set sail for Loanda, Portuguese West Africa [Luanda, Angola] for Sinclair Oil Company.

Like many geologists, he was careful about minute details (although spelling was not one of them). "After a rather uncomfortable trip across I arrived in Lisbon, September 29," he wrote Chester Washburne, New York geologist. "The Fabre liner 'Roma' is a miserable tub, very dirty and very crowded. The crowd not being members of the best families. We had 39 first cabin passengers with a cabin space for about 29...16 were either missionaries, priests or nuns. The principal cargo was about 10,000 Portuguese immigrants returning to the Azores. In fact, the boats of that line are immigrant ships, very little space being used for first cabin passengers...

"My stay in Lisbon was a short one...It so happened that a boat conveying military prisoners to Loanda was leaving October 1st, and, as it was an extraordinary trip, the office was able to obtain passage on it for me.

"Such a short stay gave me little opportunity to hunt for the articles I needed to complete my equiptment [sic]. Fortunately, the day of my arrival I had spent the afternoon driving all over Lisbon looking for helmets and tin boxes. By extraordinary luck I was able to get a couple of imitation helmets and two very good steel boxes. May I say for your information, that any men that may come out in the future should leave the States thoroughly equipted [sic]. No expectations should be entertained of being able to purchase articles of any class either in Lisbon or Loanda. In Loanda especially there is absolutely nothing in the shops. There is not a helmet in town, little or no cloth and, in fact, when one goes into a store it is with the expectation of not finding what he wants...

"After a quiet and pleasant trip of 17 days I arrived in Loanda. The trip required just one month from New York to Loanda, which was not bad."

White found the climate excellent and the men who had arrived ahead of him "most delightful fellows."

"On my arrival in Loanda I met Brandao de Mello, Beebe, Baldwin, Harp and some of the drillers who had arrived by previous boat...I think the Company is unusually fortunate in being able to secure such a fine group of men. Brandao de Mello is a delightful gentleman, who is so decent that one forgets that he is Portuguese. Beebe is a fine, agreeable chap whom it is very easy to like. Baldwin is a serious appearing young fellow, very quiet, for whom I immediately formed a fancy as soon as I met him. Harp is most agreeable, likable and friendly."

Once the amenities were taken care of, White got down to business. "I certainly consider it desirable to spend at least a month in the field with [Beebe], so that the knowledge he has gathered, which cannot be committed to paper, will not be lost to the Company...We shall leave for the field just as soon as porters arrive, which should be about the middle of November."

On Christmas day, White wrote, "From accounts of Loanda before the war you would hardly know the place. There are a bit more improvements, though not a very big bit, an extra movie house or so, and instead of things being very cheap, they are at prices that stagger one even comparing them with New York. The exchange rate goes up but the prices go up in direct proportion. The escudo the other day was $3.20 to the dollar, or had a value of about 30 cents...At present it seems that rules and regulations engendered by the war still carry on, so that it is far from easy to get anything done where the government is concerned. For guns especially, it is hard to get permits. However, even with the difficulty, de Mello seems to be able to get things by and have them ready in a reasonable length of time."

Living conditions were less than tolerable. "Hotels here are now very poor...the food was practically impossible...I spent the first three weeks at the Hotel Central, and had a room next to a roulette wheel which allowed me to sleep about five hours a night. During that time I would go up to Beebe's every day or so and get a square meal."

Rates at the hotels were steep, and they charged the same amount when the men were in the bush as in the hotel. "I finally managed to get a room in a private house and took my meals at Beebe's. The room was on the ground floor and not very pleasant but I got a good night's sleep and good food."

Houses were hard to find, but White finally managed to rent a house through Major de Mello. "We are quite comfortable up on the hill, with a wonderful view of the harbor which never grows tiresome," he wrote. "The house rent is 30 escudos or less than $10.00 real money; this seems rather incredible after American rents, does it not?"

Helmets, cloth and boxes were not the only equipment problems. White cabled Washburne, "IMPERATIVE SEND IMMEDIATELY PLANE TABLE ALIDADE PAPER." In a letter following, he noted, "When Beebe goes south, he will take this instrument with him which will leave Baldwin and myself without any."

After a quick once-over of the situation, White felt that two field parties were necessary and that they had to be particular in choosing the men. Unmarried men with "as few ties as possible back in the States" were preferable. "Men who have strong family ties, or a sweetheart they are expecting to marry as soon as they return, are more or less discontented and always longing to return. This becomes an important factor, especially when they have a touch of fever or when things go rotten, as they frequently do in foreign work. Also, the lack of female association makes them moody and restless."

Field work was not glamorous; living conditions in the camps were often confining, catch-as-catch-could in the field. It was often tough going, and opinions sometimes differed.

By April 1920, the new had worn off the friendships, and differences of opinion and personality had arisen. In August, White wrote his mother, "Things have finally come out alright. It was about time for…a good many times [I] expected to take the next boat for the States. Some times, however, it is best to sit tight and let things take their own course, which is what I did here."

White was appointed chief geologist with five subordinates. There was also talk of increasing his zone of activity to assistant manager as well as chief geologist. White attributed his new position to a number of factors—his year of residence, his "fair knowledge of the Colony," and his "fairly creditable work in a region that every one else had passed up as impossible, because of the thick brush, inaccessibility and general cussedness of it."

His new position increased his enthusiasm. "It is quite an important operation, we are just getting started and now there are between forty and fifty Americans on the job; should we get oil, it will be some big undertaking."

In October, he wrote his mother once again. "It takes a long time for a letter to travel from here to the States, so that it will be Xmas time when it reaches you.…This year we are all expecting to finish our field work and reach Loanda for that day. There will undoubtably be a big celebration. Afterwards we will remain in Loanda for office work for the next three months untill [sic] the hot weather is over, and again take to the field for the next year's work. There are seven of us including myself, and I am promised two more men for next year. They are all very high class men so we have a pleasant group.

"I am now at this little railroad station [waiting along with two men] that arrived in the last boat. We are hung high and dry, waiting for Negro cargo carriers that were promised for last Sunday. I made a trip to Malange last week and though all was arranged…the law, of the general cussedness of things, is at work and they have not arrived. Malange is a fair size town at the end of the railroad 300 miles from the coast and it requires two days to make the trip. If [the bearers] do not arrive by next Sunday I shall have to go there and stir things up. It is a long tiresome trip, only one passenger train weekly, so there is no opportunity before that time. Because of the train service, I was forced to remain eight days in Malange; the morning following my arrival our agent started by automobile to visit the district to attempt to get the necessary men and came back five days later without a man. Then there were three feast days in which no one work[s], and the government offices were closed so I could do nothing. The last day of the festivities I grabbed the chief of the government service, as I had to leave the next day; tackled him as he came from the governor's reception and had my say so I could take the train the next morning. I then left it to our agent, who does not seem to have been successful. Delays are about the most prevalent thing that occur in this country."

White was getting ready to drill in a region called [Qecisama? Qeusama?]. "The region is interesting and the people especially so. They are brown rather than black and resemble Indians more than Negroes. As their country is very dry, during the short rainy season they collect the water and store it in trees that they hollow out. Some of the trees grow to 12 to 15 feet in diameter. Formally [formerly] they were a well populated country having a half a million people, then [through] sleeping sickness and wars they have been reduced, until today, in the region where once dwelt five hundred thousand only five thousand remain…The women…are strong and able to carry a load equal to a man's from other parts…They have the custom to weave beads into their hair, which is also dressed with palm oil and a powder made from a red wood…In some of the deserted villages evidence of stone carving of a very high character is to be seen…They burry [sic] their dead in the ground and pile huge piles of stones over the place, with the result that they remain for a very long time. One could very appropriately call it the land of the dead, as there are few signs of the living."

Dangers of the Territory

Mail strikes in Lisbon made postal communications uncertain, and many of the communiques from the states never made it through to the party. White wrote early in December, on receipt of Washburne's September letters.

"I am [at] present just recovering from a good stiff jolt of fever," White wrote. "It only lasted a day, but my stomach was in bad shape so I was rather weak when the fever came. It will probably be another week before I can take [to] the field. The illness was my own fault; I had not taken the proper precautions, overestimating the healthfullness of the climate."

Despite his problems, White turned down the opportunity to return to Mexico. Chester Washburne had written that "E.B. Hopkins is looking for a good man for Mexico. I convinced him you would be the best man if available...Conditions in Mexico have become fairly good. Bandits are very scarce now. I would not hesitate on that account." White declined, explaining, "In not accepting your Mexican offer I did what may appear to be a foolish thing, as the job this year only carries a salary of ten thousand dollars; but, it is a big job, on a big concession which you know has excellent possibilities; there is a big opportunity to do a great deal of excellent work. Also, as you know all my work for the last several years has been far in advance of the drill, and I needed the opportunity to compare the development with the geological conclusions derived before drilling. I think I needed the experience, as I was getting to[o] far away from the practical side."

White spent the year searching out geological and well sites. On December 22, 1921, he wrote his mother as he sat "in a small railroad station, in a small railroad hotel with fifteen various and different odious smells waiting for a train that seems never to arrive. It is my third day of waiting and my patience is about exhausted. We are also having a nice hot spell, last night under the mosquito curtains I was bathed in perspiration for the entire night and slept rotten. This is the first hot night I have had in Africa. Today it is cloudy and cooler and there is a breeze blowing.

"This has been a year of unusually heavy rains, and transportation has broken down all over the country...I went south as I wrote you in my last letter, finished my work, which was to pick out a well site, in two days. After considerable difficulty secured an automobile whose driver was willing to try to get thru, as it was still raining at that time...Our trip was about 400 kilometers; we made two-thirds of it and then stuck in the mud. Fortunately we were near a military post who furnished men to get the machine out of the mud. At turn back, the driver charged me an awful price, and I proceed[ed] to complete the journey by [foot?] with carriers. I arrived here the evening of three days ago, and have been here ever since waiting for a train that never comes. There is a freight train promised for tomorrow but nothing is certain. The grass grows with enormous rapidity and very thick with the past two months rain...It has grown so fast and thick that trains cannot run...The grass gets onto the rails with the same effect as if they had been soaped. The result is the weekly express, which should arrive tomorrow from Malange on its way to Loanda, has not as yet reached Malange, so that it [will] probably not reach Loanda before Sunday or Monday though it is due Thursday."

White's problems centered around men and machines. Others who followed him found problems with the native wildlife as well. In the early 1920s, Sinclair Oil Company sent

G.L. Harrington worked at plane table on the geological survey west of Comodoro Rivadavia, Argentina, in 1921. R.H. Dott.

Thornton Davis to Angola under Bill Heroy. "His work was in the primitive territory and field parties had to be prepared for almost anything," wrote Wallace Thompson. "On one occasion he killed a Cape buffalo, acknowledged one of the most dangerous animals in the country. He sent a picture of his trophy to the New York office to show some of the difficulties he had to encounter. The reply from the New York office was terse... 'WONDERFUL TROPHY, NOW SET UP YOUR ALIDADE AND SHOOT IN SOME OUTCROPS.' "

A Fiery Furnace

While many geologists and U.S. oil companies were flocking to South America, British and European companies dominated the Middle East. Europeans were much more alert than Americans in developing foreign supplies, and the principal oil districts outside the United States were largely under British, Dutch and French control. Royal Dutch/Shell was the world's largest producer of crude oil. The first geological maps of northern Iraq were made by E.H. Pascoe of the Geological Survey of India, in the winter of 1918-19. The five-months' tour was hampered by persistently bad weather, and two of the camels died as a result. Pascoe did his mapping on the new Army 1-inch = 1 mile topographic sheets, and sketched cross sections of several faulted folds. He managed to cover 100 miles of territory on both sides of the Tigris River, despite the weather.

Arthur H. Noble and R.du B. Evans moved in for Shell in 1919-20. One of the areas that Noble spotted for drilling was Kirkuk. They had tried to visit it in the summer of 1919, but it was then the scene of a Kurdish uprising by Sheikh Mahmoud. G.M. Lees, then Assistant Political Officer at Hlabja, had been forced to beat a hasty retreat on horseback. Noble made another attempt to visit Kirkuk but was not allowed to do so. "It was one of the few places barred to us, as British troops had suffered losses from Kurdish raiding parties. This time I was allowed to visit the seepage area of Baba Gurgur (reported to be the Burning Fiery Furnace in which Shadrack and his companions were put); I saw enough to realize that the area was of great importance."

In 1919, Noble moved into Iraq, which had recently been taken from the Turks. "In the south, civil government had been set up as the Army moved forward," Noble wrote. "The Arabs of the Delta had accepted the new regime fairly easily, but in the religious centers of Karbela and Nejaf, stern measures had to be taken before they submitted." In the Mosul area, the population was largely Kurdish, "turbulent by nature and although they disliked the Turk, they did not take kindly to his conquerer.

Yet in the summer of 1919 order had been established in the northern provinces. There were sporadic revolts and a number of British officers were killed, but there were enough troops to garrison the important towns and to enforce British rule. As the Army was reduced and replaced by native police, everyone realized that trouble might break out in the centres of discontent."

Noble and his men concentrated their exploration in the hill country north of the Jebel Hamrin, an area well known for oil seepages, using the Army topographical maps. They first went to Qaiyarah, about 35 miles south of Mosul. "We had no transport other than what we could borrow from an Army unit stationed there, so we were confined to the vicinity of Quiyarah."

Evans managed to contrive a plane table with the help of an Army workshop, but the work was handicapped by lack of transport. "New cars were unobtainable and though the Army helped us the cars they put at our disposal were old and war-worn; therefore we had to confine our work to the more accessible places."

Noble and Evans struck out for the Euphrates Valley in July, one of the hottest months. "We found the only way to work was to start at dawn and get back to shelter at midday," he wrote. "No one could face long days in this sort of heat; it was normally 110 to 120 [degrees] in the shade, and in the sun it was blistering. The excellent Army tent was almost unbearable until the sun went down, and life was made the more difficult to bear by the flies that took shelter within."

They investigated the town of Hit, known for its great pitch lakes. "The small and squalid town...is one of the foulest smelling places I know—it is said that it can be smelt from planes. The sulphur springs (in which Job cured his boils) give off an overpowering smell of rotten eggs."

Evans quickly fell victim to dysentery and was unable to work in the hot country. He stayed with a political officer in the mountain country north of Mosul until he became stronger. But by the time he returned, the district was off limits, as it had grown unsafe.

In September, Noble's wife joined them in Mosul. "There was an abundance of game—partridge, grouse, duck and two kinds of bustard—very good to eat. With a gun in the car we could always provide our mess with game."

As the hot season passed, days cooled and nights were cold. "When on our longer journeys into Kurdistan, we slept at the houses of the headmen or aghans, as they were called; they were always hospitable and outwardly friendly. Our chief trouble was to make ourselves understood, as they spoke Kurdish and some Turkish while we spoke only a little Arabic...

"In the most remote part of the province we were entertained by a Kurdish officer of Levies, who had been trained in Turkey and spoke French. He insisted on accompanying us on our work with his troop of Levy cavalry, who wore full Kurdish dress, mounted on small but sturdy ponies and armed to the teeth. We examined outcrops to the roll of drums and with a mass of horses champing round us. I noticed that when we collected rock specimens, some of the escort also pocketed stones from the same place. I'm sure they thought we were looking for gold."

Weeks had seen the ethic at work late one afternoon when he stumbled into a clearing in the jungle and found the Resident Forest Commissioner for the area dining all alone in his tent, impeccably dressed and with proper dining utensils; yet for all he knew, there was not another European within a hundred miles.

Although far removed from the British islands, India held more promise of romance for unattached young men than most people realized. "In those long-ago days, young British ladies of good family often visited in the dominions, in an endless round of teas, dinner parties and dances," Lewis Weeks wrote. "Since the British boys, who held the Empire together, could not be furloughed home until the end of a tour of duty, aunts, uncles, godparents and friends found infinite opportunities to bring the girls to them. When the Empire was in its prime, an English boy in his early twenties might be in sole charge of keeping law and order in a territory of 10 million souls, with few other males of his race and language around and no females between the pinafore and dowager stages. The sight of a beautiful young girl from back home, chaperoned, was as thrilling as it was challenging."

On just such an occasion, Weeks met his first wife, Una Austin. Their honeymoon trip was combined with a six-month geologic survey. "Our party consisted of Una and myself, an assistant geologist, a guide and our bearer, or personal servant," Weeks explained. "There were also a Mohammedan cook who could pull five-course meals out of a tiny cook fire set in a hole in the ground (rather like a jinni rubbing his lamp), a boy to bring us bath water from the nearest stream, and another to wash the dishes. Our bearer was a Gurkha, a member of the tribe that provided the British in two world wars with some of their finest soldiers...The coolies, each loaded with 40 pounds of equipment, were primitive Naga tribesmen, who wore a narrow breechclout held up at the waist with a bit of string.

"Each day, a runner who ran along the jungle paths shaking a bell to scare away tigers and other wild animals brought us our letters, a copy of the most recent newspaper and a loaf of bread. Our Naga tribesmen were meat eaters, unlike the Hindus in our party; and once a week, if I did not shoot a deer, we had to send a servant to buy a goat in the nearest native bazaar, often a day's trip away. Sometimes he came back leading a cow...the very idea of slaughtering it raises religious scruples among Hindus. When I saw him coming with a cow, I quickly diverted him into the jungle to dispose of it quietly."

"Lewis, his helper and I have ridden our ponies 22 miles in the past two days," Una wrote her father. "We look like tramps in our baggy khaki clothes, and I feel like Dick Whittington in my boots caked with mud. About us the frogs croak, the mosquitoes hum tunes and the jackals screech. The first night, the coolies fell behind and did not bring us our bedding before dark, so Lewis and I slept in a forest rest house on a narrow string cot, which the Indians call a charpoy. For blankets to keep off the mosquitoes, we used newspaper." Mosquitoes, however, were less troublesome than leeches. "Lewis' work," Una wrote her father, consists of "wading down into rivers and mushy marshes gathering leeches...in search of those funny little anticlines."

**Since there were no paths cut through the jungles of northeastern India, the best way to travel was by elephant.
L.G. Weeks/AAPG.**

reduced and replaced by native police, everyone realized that trouble might break out in the centres of discontent."

Noble and his men concentrated their exploration in the hill country north of the Jebel Hamrin, an area well known for oil seepages, using the Army topographical maps. They first went to Qaiyarah, about 35 miles south of Mosul. "We had no transport other than what we could borrow from an Army unit stationed there, so we were confined to the vicinity of Quiyarah."

Evans managed to contrive a plane table with the help of an Army workshop, but the work was handicapped by lack of transport. "New cars were unobtainable and though the Army helped us the cars they put at our disposal were old and war-worn; therefore we had to confine our work to the more accessible places."

Noble and Evans struck out for the Euphrates Valley in July, one of the hottest months. "We found the only way to work was to start at dawn and get back to shelter at midday," he wrote. "No one could face long days in this sort of heat; it was normally 110 to 120 [degrees] in the shade, and in the sun it was blistering. The excellent Army tent was almost unbearable until the sun went down, and life was made the more difficult to bear by the flies that took shelter within."

They investigated the town of Hit, known for its great pitch lakes. "The small and squalid town…is one of the foulest smelling places I know—it is said that it can be smelt from planes. The sulphur springs (in which Job cured his boils) give off an overpowering smell of rotten eggs."

Evans quickly fell victim to dysentery and was unable to work in the hot country. He stayed with a political officer in the mountain country north of Mosul until he became stronger. But by the time he returned, the district was off limits, as it had grown unsafe.

In September, Noble's wife joined them in Mosul. "There was an abundance of game—partridge, grouse, duck and two kinds of bustard—very good to eat. With a gun in the car we could always provide our mess with game."

As the hot season passed, days cooled and nights were cold. "When on our longer journeys into Kurdistan, we slept at the houses of the headmen or aghans, as they were called; they were always hospitable and outwardly friendly. Our chief trouble was to make ourselves understood, as they spoke Kurdish and some Turkish while we spoke only a little Arabic…

"In the most remote part of the province we were entertained by a Kurdish officer of Levies, who had been trained in Turkey and spoke French. He insisted on accompanying us on our work with his troop of Levy cavalry, who wore full Kurdish dress, mounted on small but sturdy ponies and armed to the teeth. We examined outcrops to the roll of drums and with a mass of horses champing round us. I noticed that when we collected rock specimens, some of the escort also pocketed stones from the same place. I'm sure they thought we were looking for gold."

When Noble was ready to return to Baghdad, trouble had broken out among the tribes in the desert and the road was unsafe. "All cars had to go in convoy with an escort and that meant an unpleasant journey," he noted. He decided to try the river instead. "There were no river steamers or launches as the rapids of the Fathah gorge...are too shallow, but keleks—rafts of inflated pigskins—often made the journey from Mosul to Baghdad. I had a kelek made, large enough to take the three of us, two servants and two boatmen."

The raft had two little houses of palm leaf built on the deck. "We had a great send off from Mosul; hundreds of people came to the river bank to see us start. At the last moment an aged man wearing the turban of a Hadji climbed on board. We protested but it was no good; after polite greetings he settled down on deck and our well-wishers assured us that to have a Hadji with us would assure a safe journey. He was a nice old man, but he prayed loudly and raucously at all hours of the day and night."

They made rapid progress, for the Tigris was swift with currents from melting snow in the mountains. "The trip lasted three days and an easier or more restful mode of travel would be hard to find. We left our Hadji at Samarrah, as he was a priest of the famous mosque with the gilt dome."

Noble spent the summer in Baghdad making maps and writing reports. Temperatures in the shade often went up to 120°, but were relatively cool in the cellar of a Turkish-built house. "The upper rooms were impossible to live in," he wrote. "Metal objects in them got so hot that one had to pick them up gingerly. Everyone in Baghdad slept on the roof, and the nights weren't too bad except when a dust storm blew up and then it was hell."

Noble's wife was "a competent draughtsman and undertook the drawing and painting of the maps—no easy task in a climate that made one drip at the least exertion. I'm sure few explorations...presented maps of the quality of ours."

By late summer of 1920 the revolt had spread throughout the Delta and much of the country north of Baghdad was out of control. "It was clear that the country wouldn't be safe for a long time," Noble wrote. "I, therefore, put the situation to London and got permission to return...We got home in time for Christmas."

Meanwhile, D. Dale Condit, Ralph Howell and several assistants surveyed the Baluchistan area for Whitehall Petroleum, a British Company. Condit was an inveterate investigator. "In the field he set a pace few could maintain," one friend wrote. "No obstacle deterred him. He would swim swiftly flowing rivers with his horse, wade through crocodile infested swamps, cross dangerous rock slides, descend old shafts and crawl into abandoned mines occupied by snakes and wild animals. He never spared himself even when repeatedly incapacitated by malaria and other tropical diseases, and he was ever solicitious for the well-being of his associates, juniors and seniors alike."

During the survey, an uprising occurred in a native village. Howell was ambushed by Marri tribesmen and killed.

Asiatic climates ranged from the tropics to the mountain ranges, and geologists had to be prepared for extreme changes. When Standard of New York (Socony) hired Lucian Walker and three others to look for oil prospects in Inner Mongolia, the expedition made its way to Pastow [Pastowchan or Shi-Pastow], a city of 100,000, five miles north of the north bend of the Yellow River and 100 miles west of the west end of the railroad. The climate was bitterly cold, but the men quickly adopted native wear. When one member of the group snapped a picture of Walker for his friends at home, Walker was bundled in a coat and pants of green goat skin lined with camel's hair, and Mongolian felt boots.

Proper Tea and Water-Buffalo Milk

Sir Weetman Pearson (Lord Cowdray) had been active in the Mexican oil fields prior to the war, but on the advice of his geologists, he sold out and moved to new territories. This time it was the jungles of the Brahmaputra and Surmay valleys, the adjoining hill tracts in Assam, eastern India, and the desert plains and adjoining hills of Punjab and Baluchistan in what is now Pakistan. He set up company headquarters in Shillong.

"Life at a hill station in British India in the early 1920's was still delightful," wrote Lewis Weeks. "Our bachelor quarters were pleasant. Each of us had his bearer, or man servant, who laid out his clothes, drew his bath and brought him his tea each morning. Tea was served in bed with semi-refined brown sugar and water-buffalo milk, first boiled to be safe. And instead of butter, toast was spread with ghee, refined butter oil which requires no refrigeration. My bearer even insisted on putting on my slippers...

"To a young American from the farm country, I must say it was pleasant. In the early morning while the dew was still on the grass, I rode my Arab pony on paper chases. Even where there are no foxes and no hounds, the British still managed to hunt. Two or three people would go ahead to scatter bits of paper, making sure the clues were not too frequent or too obvious. All the riders congregated at the starting point and the first one to find the last bit of paper in effect received the brush of the nonexistent fox.

"Once a week I played polo with British army officers, civil servants and the native rajahs and their sons. And I also rode my ponies in the yearly gymkhana races. A gymkhana was a club organized for sports, usually with...dining room and bar, rather like a country club in the United States. Nearly every afternoon there was some kind of sports or social event, following which everybody put on black tie and tuxedo for dinner with the ladies. The British tradition of dressing for dinner while stationed abroad probably served two purposes. It exhibited the British determination to hold on to convention and civilization even amid the most seductive of foreign environments, where it might be all too easy for a proper gentleman to let himself go. And it undoubtedly served to intimidate the natives. Which purpose came first, I never could figure out."

When Lewis G. Weeks surveyed India in the early 1920s, polo was an important gentleman's sport at Shillong, Khasia Hills, India. Polo was played during the hot and monsoon seasons, and both civilians and military had their own teams. L.G. Weeks/AAPG.

Weeks had seen the ethic at work late one afternoon when he stumbled into a clearing in the jungle and found the Resident Forest Commissioner for the area dining all alone in his tent, impeccably dressed and with proper dining utensils; yet for all he knew, there was not another European within a hundred miles.

Although far removed from the British islands, India held more promise of romance for unattached young men than most people realized. "In those long-ago days, young British ladies of good family often visited in the dominions, in an endless round of teas, dinner parties and dances," Lewis Weeks wrote. "Since the British boys, who held the Empire together, could not be furloughed home until the end of a tour of duty, aunts, uncles, godparents and friends found infinite opportunities to bring the girls to them. When the Empire was in its prime, an English boy in his early twenties might be in sole charge of keeping law and order in a territory of 10 million souls, with few other males of his race and language around and no females between the pinafore and dowager stages. The sight of a beautiful young girl from back home, chaperoned, was as thrilling as it was challenging."

On just such an occasion, Weeks met his first wife, Una Austin. Their honeymoon trip was combined with a six-month geologic survey. "Our party consisted of Una and myself, an assistant geologist, a guide and our bearer, or personal servant," Weeks explained. "There were also a Mohammedan cook who could pull five-course meals out of a tiny cook fire set in a hole in the ground (rather like a jinni rubbing his lamp), a boy to bring us bath water from the nearest stream, and another to wash the dishes. Our bearer was a Gurkha, a member of the tribe that provided the British in two world wars with some of their finest soldiers...The coolies, each loaded with 40 pounds of equipment, were primitive Naga tribesmen, who wore a narrow breechclout held up at the waist with a bit of string.

"Each day, a runner who ran along the jungle paths shaking a bell to scare away tigers and other wild animals brought us our letters, a copy of the most recent newspaper and a loaf of bread. Our Naga tribesmen were meat eaters, unlike the Hindus in our party; and once a week, if I did not shoot a deer, we had to send a servant to buy a goat in the nearest native bazaar, often a day's trip away. Sometimes he came back leading a cow...the very idea of slaughtering it raises religious scruples among Hindus. When I saw him coming with a cow, I quickly diverted him into the jungle to dispose of it quietly."

"Lewis, his helper and I have ridden our ponies 22 miles in the past two days," Una wrote her father. "We look like tramps in our baggy khaki clothes, and I feel like Dick Whittington in my boots caked with mud. About us the frogs croak, the mosquitoes hum tunes and the jackals screech. The first night, the coolies fell behind and did not bring us our bedding before dark, so Lewis and I slept in a forest rest house on a narrow string cot, which the Indians call a charpoy. For blankets to keep off the mosquitoes, we used newspaper." Mosquitoes, however, were less troublesome than leeches. "Lewis' work," Una wrote her father, consists of "wading down into rivers and mushy marshes gathering leeches...in search of those funny little anticlines."

Since there were no paths cut through the jungles of northeastern India, the best way to travel was by elephant.
L.G. Weeks/AAPG.

A vine-rope bridge (jhula) high above Sach Pass in the Himalayas crosses above the Indus River. "The jhulas are made of vines," Weeks wrote. "These are braided to make three thick ropes. The main rope is that on which the traveller walks, and the two side ropes are used to maintain balance. L.G. Weeks/AAPG.

Lewis Weeks's geological investigations took him high into the mountain ranges in 1923. The photo was taken at 17,500 feet in the high Himalayas of Chamba state. The Guardhars range in the distance was 22,000 feet. Weeks camped at 17,000 feet and actually climbed to 19,000 feet at one point. L.G. Weeks/AAPG.

Lewis Weeks's expedition into the Indian jungles of Assam in 1922 included 40 Angami Naga tribesmen recruited from the Naga Hills for "Coolie duty." Weeks shot a wild boar, and "there was good meat for all that night."
L.G. Weeks/AAPG.

Una stayed with Lewis even when he moved into the heart of the jungle. "The men arise early to go out to work," she wrote, "and as they do not return until tea time, I eat my lunch alone, served in style by our bearer...off an enamel plate. When I return to Calcutta, I shall have to rehearse my manners. I eat most things with my fingers and throw anything I disapprove of on the ground. Jumman-Singh acts as our personal valet, makes our beds, mends and washes our clothes and handles the money for us. Poor as these people are, it's a point of honor with him to be utterly scrupulous in handling our money, and we find we can trust him with any amount...

"The cook is marvelous, too...Occasionally, in the early dawn, Lewis will go out and shoot a wild pigeon and bring it to me saying, 'There's your breakfast.' Other times, we eat a lot of the scrawny little jungle chickens bought live from the hill people. I hide my eyes when the necks are twisted...

"The Naga hill people wear their hair in a curious fashion. It is cut very short and straight at the bottom of the head and left long on the top. This gives a cap effect. They have large holes bored in their ear lobes, which they keep open with soiled pieces of rag. Yesterday, the village headman presented me with a leg of a deer and twelve oranges which he extracted from his not-too-clean pockets. Later I received a gift of a large papaya, some sugar cane which is much appreciated here, rather like candy with us, and a freshly slaughtered pig...

"Yesterday the postman brought me a wedding gift from one of our friends, a fan in the latest Paris mode with eight gorgeous ostrich plumes caught up in an ivory handle. In my khaki pants and riding boots, with my face all broken out in mosquito bites, I looked at myself in the mirror over the wash basin and tried to imagine how I will look when I get a chance to carry the fan to a party."

Lewis Weeks was a young man when he explored India in the early 1920s. His new bride accompanied him for most of the trip. L.G. Weeks/AAPG.

Although the accounts of frontier petroleum exploration conjure up images of Indiana Jones or James Bond, the actual story of exploration outside the confines of urban civilization was more one of hard work and methodical investigation.

Robert H. Dott, Sr., was hired by Standard Oil of New Jersey as a cub member of "Jersey's" first reconnaissance team into Bolivia and Argentina in 1920–1922. The following section is illustrated with Dott's personal photographs and comprises excerpts from his diary, and from letters to his father, describing details of his routine and of the authentic adventures encountered in working a sparsely settled region.

The party, consisting of J. B. Mertie, Jr., Harvey Bassler, D. S. Birkett, and R. H. Dott, and led by Eugene Stebinger, sailed from New York on Memorial Day 1920, changed ships at the Panama Canal, and anchored off Mollendo, Peru, June 27.

"Had to be swung ashore in a chair." Dott noted in his diary.

June 28. "After much trouble, we got away at 12:00. Had a delightful ride to Arequipa..." June 29. "Arequipa is a beautiful place with a wonderful climate. Haven't gotten our baggage yet." June 30. Bassler and I took a horse-back ride in afternoon. I got pretty sore." July 1. "Cannot get our baggage. No change of clothes in a week—we're a fine bunch of bums."

They finally retrieved their baggage and on July 2, set out by train for La Paz, Bolivia, *via* a boat across Lake Titicaca. They arrived in the Bolivian capitol at 12:30, July 3.

The party split. Stebinger and Mertie were to go as far as Santa Cruz together, partly by rail, the rest by riding mules and pack train. There, Mertie would continue across the Chaco to the Paraguay River, travelling by boat to Buenos Aires and by rail to Salta. Stebinger would go south on the Santa Cruz-Yacuiba cart road, east of, and parallel to, the front Subandean range; they left La Paz July 7.

Bassler and Dott left July 9, for Yacuiba, *via* Juyjuy and Salta, Argentina.

The Bolivian leg of the journey was across the *Alto Plano* (high plains, elevation 12,000 feet). A railroad was under construction to the Argentine border and had reached Atocha, with 20% yet to go.

"...Atocha sure is a bleak, dirty place," Dott recorded in his diary on July 10; on July 11, "Left Atocha at 10:30 a.m. via auto. [a high-wheeled open Mercedes with the top down.] Had a very dirty, windy ride to Tupiza. Got there at 3:30...This is rather a clean-looking town." The next morning they left by auto for the border town of Villazon, where they cleared customs and immigration, and crossed to la Quiaca, Argentina. At a lunch stop on the way they learned about the revolution in La Paz. Dott recorded July 13: "La Quiaca is a rotten town. Built on a desert, wind 40 miles per hour, much sand. Elevation 10,000 feet." Their large baggage—camping equipment, saddles, etc.—had come from Atocha on 2-wheel mule carts; also, they expected to be met by a Dane named Hoppe—probably an Argentine citizen—who was to help them in any way possible—through customs, acquiring animals and peons, etc. They had hired a young Bolivian—Angel Barrientos—who had attended a Methodist School in La Paz, and spoke fluent English.

July 15. "Hoppe showed up about 10:00 p.m.; contrary to expectations, we found him a very likeable fellow." He was also fluent in English with some amusing Danish twists.

The baggage showed up about the same time Hoppe did, so July 16 they "left at 8:30, after the usual scramble over baggage and tickets...Got to Juyjuy at 7:30...Juyjuy is a clean, nice-looking town." Juyjuy has a much lower elevation than La Quiaca, and they saw oranges for the first time.

On July 17 they traveled to Salta, the provincial capitol, and arrived at 10:00 p.m. "Salta

The only way ashore at Mollendo, Peru, in June 1920 was via lift. Robert H. Dott was with the Stebinger party on its way to Bolivia for Standard Oil Company (New Jersey). R.H. Dott.

seems to be a nice city"—about 6,000 feet in elevation. After the Alto Plano, they seemed to have reached paradise. They stayed in Salta 10 days.

Dott's entry for July 28: "Got away for Embarcación...We were lucky—train off the track only once, and we got our baggage and mules." Embarcación is about 100 miles south of Yacuiba, Bolivia, at the International boundary. At that time Embarción was the end of the railroad.

Here ended their travel by rail. They had acquired six mules and a mare in Salta. Mules will stay with a mare, especially a white one, with a bell on her neck, and will follow in line behind her. This determines the rate of travel and distances covered per day, though availability of water and pasturage must be considered as well.

They also stocked up on canned food and other staples in Salta, for "living off the country" was much better in theory than in practice.

Standard Oil geologists pose on the plaza at Arequipa, Peru, en route to La Paz, Bolivia, June 30, 1920. Left to right: A Cuban passenger by the name of Varona; Robert H. Dott, Sr., Harvey Bassler and Eugene Stebinger.   R.H. Dott.

Jungle Trails

The adventure began August 3, as Dott recorded: "Started from Embarcación. Camped at Campo del Medio—about 4:00, distance 12 mi. Dog got our meat."—fresh meat purchased in Embarcación. On August 4 they traveled 30 miles and on the 5th they "Managed to buy eggs; ...arrived at Tartagal 4:30. 21 mi. There is a sort of hotel here. This is a place of about 50 population." A few years later, after the company began drilling, and found oil, Tartagal became operating headquarters for northern Argentina and, to some extent, Bolivia as well.

The next day they managed to buy more eggs and a chicken; then traveled 25 miles, to within 15 miles of Yacuiba.

Dott's entry for August 7: "Started at 8:30, and got to Yacuiba at 3:30. Celebration, and as tomorrow is Sunday, we can't get away until Monday." August 8. "The Bolivians are celebrating their Independence Day. Had 3 days—6, 7, 8 and drank much alcohol."

Having arrived in the concession area, the geologists were ready to go to work. With the option, the company had obtained geological reports, notes and maps that had been prepared for the late Richmond Levering, the grantee of the concession. Among these was a general map showing the front range and the cart road just east of it, showing the crossing points of rivers and creeks that cut through the range.

The assignment given Bassler and Dott was to "detail" the creek valleys upstream, with a pace-and-compass traverse starting from the crossing landmark, and observe and record dips and strike of the strata upstream until a reversal of dip could be seen. Also, to find, locate and sample oil seepages, if any. It was the numerous reports of seepages—oil springs—that had attracted attention to the region.

August 10. "Came to Aquarenda and on up [Aquarenda] Quebrada and into another, small clearing owned by Epifameo [Yaluin], and camped. He came along to show us an oil seep." August 11. "Visited oil seep with Yaluin. Got some samples. He told us many wild tales about wild animals—too much alcohol."

August 12. "Rode toward Carapari...West over a very steep trail; climbed 3,400 feet. August 13. "Came back to Yacuiba. Expected to find Hoppe, but he had not come.

On August 14, (writing from Yacuiba) Dott told his father, in part:

"I wrote you some time ago about the revolution in Bolivia. A fellow (Col. R. Obando) who lives in Caiza, a small town a short distance north of here, has been appointed minister of War. He is to be in town today, and they are making elaborate preparations for his entertainment. The band is now practicing but I am afraid they won't be ready. Here at the hotel yesterday they killed two turkeys, three chickens and a duck. Now they are working on two cows' heads.

"With regard to the revolution here, there have been nine deaths and 17 wounded, in the whole country. One regiment in Sucre was all that resisted and that was the only place there was any fighting. One man who had made himself very unpopular in La Paz was shot

Standard Oil geologists investigate the Peima oil seepage in the foothills of the Andes Mountains in 1920. R.H. Dott.

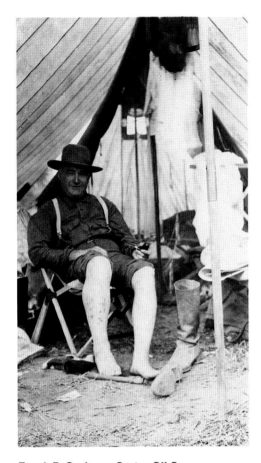

Frank P. Graham, Carter Oil Co. production man, went along "to look over the country" with the Standard geologists in southern Bolivia in 1920. Graham displays the sores that resulted from scratching the small blood-blister bites of a sand fly. "Oh how they itched!" Dott recalled. R.H. Dott.

in bed. On the whole, it was more of a hand-over than a revolution. They took the president with 30 cadets, whereas he had 40 armed guards in his residence, who made no resistance at all."

During that time they were joined by Frank P. Graham, an experienced oil production man from Oklahoma with Carter Oil Company, a Standard Oil Company (N.J.) subsidiary. He was to accompany the geologists to assess the nature of the terrain and difficulties it would cause for moving in equipment, operating, and the like.

Diary, August 15. "Hoppe and Graham got in at 6:30." August 17. "Went up Rio Carapari...Shot a parrot." August 19. "Tried to shoot ducks." August 21. "Drafting in camp." On August 22 the party, now including Frank Graham, started eastward along the 22° parallel toward D'Orbigny, on the Pilcomayo.

Villamontes, Bolivia
Sept. 2, 1920
"My Dear Dad–
"We've moved on to here, and are now waiting for Stebinger, who will be here tomorrow.

"Since I last wrote you, we have been east from Yacuiba to D'Orbigny where the Pilcomaya River crosses the 22-degree parallel. From there we came up the Pilcomayo to here, in quest of something of geologic interest, but none rewarded our efforts. The only things we saw of interest were some wild Indians, whose village we visited, and whose picture we took. It really was quite an experience—one that few people get to have.

"Though perfectly friendly to whites, they are completely wild, living in the same way their ancestors have lived for centuries. Their houses are made of grass, and are very small, and none too clean. They are conical in shape, about 5 feet high by 8 feet in diameter, and have a door in one side, made of grass. There were about fifteen such houses in the village, sheltered away among small trees and wild cane, along the bank of the Pilcomayo River. Their sole occupation is fishing and hunting. They are very clever fishermen, and I guess fish is their main diet. They use a net, about 10 feet long by four wide, strung on two cane poles. They walk along in the shallow, muddy water until they see a fish, then dip in the net, and up comes the fish. How they see anything in the water is more than I can figure out, for the river is just about like the Missouri.

"We got into quicksand and one of the big mules got in so far that he got wet all over, then mine started, but didn't go clear down. We finally extricated the entire outfit, and got safely out.

"We bought some of the fish, and had a big meal on them, but they weren't very good. They looked like carp.

"Birkett, the man who had been with Stebinger got in yesterday and brought us out first states' mail—letters dated June 13 and 14, but looked mighty good, anyway.

"Villamontes and the land around here is owned by Germans. We camped two nights at

Germans' places, coming up, and they couldn't be nice enough to us. They gave us meat, bread, vegetables and pasture, and wouldn't take a cent. They are hard workers and, I guess, are making lots of money. The whole outfit is under one company—Stout and Co., who have a store here—similar to Leach Bros. They raise cattle mainly, which they have to drive to Embarcación—a distance of nearly 200 miles."

Sept. 6, 1920
"Since I started this letter, much has happened...The man who was with Stebinger got in, and Stebinger came...[September 3]...he decided to have all of us go to Salta, to draw up the maps for him to take to N.Y."

Dott's diary entry for September 4: "(Stebinger) changed his mind, damn it, so I'm stuck up here with Bassler for 2 more months."

Dott's letter of September 2 (from Villamontes), completed September 6 (from Caigua Creek camp), continues: "Stebinger announced that Bassler and I should stay on here, in order to get a little more work done, so now we are in Caigua Creek, about 3 leagues north of Villamontes. There is an oil seep 3 leagues [1 league = 5 kilometers = 3 miles, approx.] in from here, and we are going there tomorrow. We can ride about one-third the distance, then must leave our mules, as there are six deep pools where we have to swim. We are planning to leave early in the a.m., with camp equipment. Have hired a guide, and three Indians to carry our stuff. We hope to have to spend only one night there. From the weather tonight, we're not going to have any too nice a day, either, so that all in all, it will be a unique experience.

"We have annexed a new member to our party, a German cook [Ernesto]. Now we have quite a cosmopolitan party—two Americans, two Argentinians, one Bolivian and one German.

"I am in hopes that Stebinger decides to put me with Mertie, as he is a much nicer fellow than Bassler, although Bassler has been far more human lately.

"The plans now are for us to stay in the field until real bad weather—November—then go to Salta for office work, and wait for Stebinger to return—if he does. Then further plans will be made.

"Stebinger seemed disposed to recommend this area highly, and if they do take it over, there will be two years' work ahead, he told me. I don't know whether they expect us to stay that long, or not, and I'm not sure that I will, but we'll decide that when the time comes.

"Mr. Graham...left us at Villamontes, headed for N.Y. He was a fine chap, of the rough and ready oil-field type, with a big heart, and I sure was glad he stayed with us as long as he did..."

To anticipate a few years, Stebinger's optimism was proved justified by the discovery of several good oil fields. In 1925 Fred R. Sutton reported a fine company camp at Buena

Eugene Stebinger joined the Bassler party at Villamontes, Bolivia, September 7, 1920. R.H. Dott.

**(Following page)
Standard Oil's crew camped at the mouth of the Rio Carapari in southern Bolivia in 1920. The German cook, Ernesto, and one of the peons, Juan, posed near the tents. R.H. Dott.**

Robert Dott (on left) visited the Mataco Indians "at home" near the Rio Pilcomayo, D'Orbigny, Bolivia. They fish with a net.   R.H. Dott.

(Right)
In 1920, the Mataco Indians lived along the Rio Pilcomayo in the Chaco, Bolivia and Argentina. Many went into the foothills to harvest sugar cane each year. R.H. Dott.

Vista, 25 miles north of Villamontes.

After the adventure on Caiza Creek, Bassler and Dott moved camp to Tentaihurati Quebrada, near Camatindi Mission. From there Dott wrote his father on September 9:

"Our swimming party was quite an experience, but I was sure glad when we got back to camp. We took off all our cloths, excepting a coat, and I bought some native sandals from one of the fellows who carried our stuff. We worked for four hours in this garb, crossing and recrossing the creek and had to swim four or five deep pools. One of the natives nearly got drowned, and I guess would have if I hadn't pulled him out. We found a lot of oil and gas seeps, and a closed anticline, which looks pretty good, so our efforts were well rewarded. How they'd ever get tools in to drill is more than I can see.

"I got my camera wet in one of the pools, and now it is out of order, with water in the lens and shutter. I hope it will be O.K. when it dries.

"Tomorrow we are going up this stream we are camped on, and hope to get a good section across the anticline. The guide we have hired is afraid of tigers [jaguars], so I may get a chance to spend a few bullets."

Diary, September 15: "Went to Machareti. In Tiguipa, our train was mistaken for that of the Archbishop, and they tolled the mission bell as we went through town." September 17: "Went to Tiguipa." September 20: "Got a wire in a.m. from Stebinger to go to Embarcación. I'm to go to Peru. What's up?"

September 24: "Yacuiba again. Cold weather. Much states mail. Bought a tiger skin." September 30. "To Salta. Plans changed again. Bassler and Birkett to Peru. Mertie and I to work Argentine." October 5. "Bassler to B.A. I'm not grieving much."

Stebinger had left or soon would leave for New York, with field notes, maps, and at least oral reports from Mertie and Bassler, to report on the three-month investigation of the Levering concession. Upon reaching Salta from Villamontes he had found instructions to drop the Bolivian work for the time being for work in Argentina, and to send a man to Peru. The areas to be investigated in Argentina were Salta and Juyjuy, in the north, and Mendoza and Nuequen, in the west.

Dott's diary, October 16, 1920. "Mertie got in [from Buenos Aires] in evening. Brought cigarettes, etc.—just like Xmas."

On October 22 they went to Embarcación to begin examining the southern extension of the Bolivian front range (Sierra Aguarague), which disappears between the border and the latitude of Embarcación; they worked together until early November.

Dott's diary entry for November 7th reported: "Moved to Tartagal...Had [canned] plum pudding for supper to give Mertie a good send-off." He left the next morning to Salta.

Typical pack train of animals strung out on the trail, single file, behind the bell mare. White background rock is a massive bed of gypsum of Late Jurassic age, in southern Mendoza Province, Argentina. R. H. Dott.

Ernesto, a German, walked into the Standard Oil camp one day and declared he was a cook. His statement proved to be substantially correct. He was "very ingenious in improvising cooking facilities," according to Dott. R.H. Dott.

On Train, Salta–B.A.
December 19, 1920
"My Dear Dad:

"We got back into semi-civilization last Monday, when we completed a six-week trip to the north and northwest of Embarcación. I believe that I last wrote you from Tartagal, the night before Mertie left for the south. [Nov. 7, 1920.]

"I waited there about four days for the new man—B. H. Frasch, who came down here from Colombia and Venezuela. After he arrived, we started up the Rio Seco (Dry River), which flows between two mountain ranges, and through an uninhabited country, humanly speaking, but a country thickly inhabited by very ferocious insects, and other wild animals. It seemed to us that we were the only human beings in the world, and that all the insects, being devoid of other prey, were trying to keep alive by feeding upon us.

"The worst of these was a little fly, called "marigui", a sort of a sand fly, I think, which leaves a little blood blister as it bites. It is not particularly malignant but itches considerably, and when so many of them get around one's face that it is impossible to read the compass, they are annoying, to say the least. The mosquitoes were also present, in sufficient numbers, but we were prepared for them with an Alaska mosquito-proof tent. The only trouble was that if we'd wait for supper until the sand flies went away, the mosquitoes would prey upon us.

"Taken all-in-all, it was a rather disagreeable trip, for we had considerable rain, it was quite hot, and the bugs were awful. As I said, practically no one lives in the Rio Seco valley, although there is a trail that cattle smugglers use to bring cattle out of Bolivia, so we were able to get clear to the head of the stream and over the divide to the north. In the northern part, near the Bolivian line, more people live.

"We had one rather amusing experience—although it didn't seem quite that at the time. We made our camp near the headwaters of the Rio Seco, and took a peon, a guide and enough of a camp outfit for three days, in order to make a rather extensive trip to the north and northeast. The second night out, by a small side-trail, we got into the Rio Carapai and sent the guide and peon up the river to the mouth of a side canyon which we would have to go up in the morning. We went downstream a ways, to look at some outcrops.

"When we got back to the mouth of the creek no one was to be seen, but their tracks went up the creek. We followed until almost dark, calling all the time, but got no response so finally went back to the main river and spent the night on a sandbar with no covers but a rain coat, and nothing but a little chocolate to eat. It was too dark to find any pasture for our mules, so we tied them up and they also went hungry. There are lots of Jaguars, (Tigers, the natives called them) in that country, and mules are deathly afraid of them so we were afraid that we'd wake up in the morning without any mules, but we spent the night alternately sleeping, hunting fire wood in the dark, and all the time fighting mosquitoes and ticks. Our peon came back soon after daylight and told us the guide didn't want to stop at the mouth because there was no pasture for his horse, so they'd gone on about a league. We went on with him, and had some breakfast, then rode to camp.

"In the six weeks that I was out, there were only five days that I was not in the saddle at least part of the day, and most of the time we rode twelve hours a day so you can imagine that I was rather glad to get back to Embarcación, and am glad for a rest before starting out on another job.

"But I am two weeks behind schedule, so can't indulge myself much of a vacation. We got into Embarcación about 3:00 one afternoon, with our nine pack mules, and got the packs all ready to take to Oran the next morning. We left Embarcación at 8:00 in a box car. There had been a wreck the day before, and the regular train hadn't got in. We were taken around the wreck on a hand-car drawn by a horse. Our baggage wasn't transferred, and didn't get to Oran until the next morning. On our arrival at Oran, I found a wire from

When Robert Dott explored for Jersey Standard in southeastern Bolivia, half the battle of geological exploration was just getting into the field. The crew had to be fed, mules and horses had to be loaded, and that was only the beginning. The early morning routine could be tedious.
R.H. Dott.

King, our land man, (who has been working in the Provinces of Salta and Juyjuy, getting information on concessions) saying he would be in Embarcarción that night, so I had to go to...the junction and head him off. He had a big pile of mail with him, so I was glad to see him.

"I expect to spend ten days or so in B.A., getting an outfit together, and to spend Christmas there. Then King and I are going to Mendoza, where I have a lot of field work, and he a lot of land work to do. I sure am glad to be starting out on a new piece of work that no one else has had, or will have anything to do with. I am due back in B.A. about Febr. 1 for a conference."

Mendoza is located east of the Andes at latitude 33 degrees south, about 100 mi east of Santiago, Chile. It is in an earthquake zone and there were newspaper reports of current activity. Dott left for Mendoza December 29, 1920, and arrived the next day. "...This seems to be a nice town. Met the Co. representative [West India Oil Company]. No sign of earthquake." January 2, 1921. "Packing, etc. Felt a slight earthquake tremor in p.m.—my first experience..."

Dott first visited a small oilfield at Cerro Cacheuta 20 miles southwest of Mendoza. The field had been discovered in 1912, but was never economic. He described the experience in a letter to his father dated: Mendoza, January 13, 1921:

To R. M. Dott
1/13/21
Mendoza, Argentina

"...this country is a paradise to work in. There is practically no vegetation. All that grows, except by irrigation, is a little sage-brush. The place I went to is about 20 miles from here. We went by train to a town about half-way, where we got a buggy, or rather a surrey. In this we went to the oil field, which comprises about half a dozen small wells. I spent two days around there, and then went around the mountain to a small town on the west side. The trip was made over a little-used road, and we had a poor bunch of horses, so had a hard time. When we got to within 2 leagues of our destination, instead of keeping on going down a stream bed we were following, we turned up another, following some old tracks, and the horses refused to pull; so we saddled three of them, and started on. We got to a house and learned that there was coach road to where we wanted to go, but the fellow took us there over a pretty good trail, then went back and helped move the rig to his house. The next day we borrowed some mules from a man who owns a lot of land around this town—Potrerillos was the name of town, and had a fine ride all over the country. I came back that day on the train, and sent my interpreter back with the outfit.

"I have been spending this week in getting a lot of maps, etc., together, and expect to leave on Monday for San Raphael, to be gone two or three weeks. I have got hold of an American [Kirkpatrick] who owns a concession down that way, and who has an outfit of

At Rio Grande, near Portezuela de Viento, Mendoza province, Argentina, a helper, Juan, crosses the river on a device called a "latigo"—several wires anchored on each bank. "The river was swift, deep and cold," Dott noted. R.H. Dott.

(Opposite)
When Robert Dott explored the province of Mendoza in 1921, the crew was led by a man by the name of Kirkpatrick. Kirkpatrick not only owned the horses and camp gear but also some of the mining claims in the area. R.H. Dott.

(Below)
The ferry across the Rio Negro at Conesa, Argentina, was a simple affair, barely large enough for the geologists' automobile. R.H. Dott.

horses and mules, to go with me. We are going to travel pretty light, and I hope to get back in a short time for a conference in B.A.

"...have heard of some other Americans who are around. Today, three got in from B.A., among them...a geologist...a man named Bennett...

"My work is getting into a sort of routine, and as the novelty of it all is wearing off I don't have the inspirations that I had at first. We had a slight earthquake here a few days ago, but it did no damage. Someone predicted that Mendoza would be destroyed on the 12th, but she still stands. Yesterday we saw some ruins of two old churches which had been destroyed in 1860. Some of the walls are still standing. The walls were over 6 feet thick. I wish they'd get a real strong one before I leave, for I should like to experience a real one."

Dott wrote his father next, from San Rafael, about 125 miles from Mendoza, reached by rail.

Transportation was basic in South America. To get any supplies or equipment the 60 miles between Embarcación, Argentina and Yacuiba, Bolivia, meant sending it by two-wheel carts drawn by mules.   R.H. Dott.

The geologists supplemented their diet with food fresh-off-the-wing. Left to right: G.L. Harrington, geologist-party chief, W. Christy and George Brady, U.S. State Department, feast on prairie chicken for lunch en route from Rio Colorado to Comodoro, Argentina, in March 1921. R.H. Dott.

January 19, 1921

"I am getting ready to leave for the field again…I am waiting for Kirkpatrick's animals to get in here, so we can start. I hope they arrive pretty soon, for I want to get underway. He persuaded me that a pack train would be the best way to go, so I have abandoned the buck-board idea, temporarily at least. I am going to look over the country, and if there is any more work to be done down this way, I think I'll do it in some other way than by pack train. I go from here to the Rio Grande and back. It is about 200 miles down there, and I am figuring on 12 days for traveling, and 8 more for side trips. I want to be in B.A. by the 15 of Febr., for a conference, and also to get some dental work done.

"This trip, I am traveling very light. In Bolivia we used as many as 9 pack animals; now, we are taking only 2. I am taking only a little food and my bed, and a few clothes. I want to get over the ground as fast as possible, and that seems to be the best way of doing it. Kirkpatrick has quite a string of horses and mules, and he thinks we should be able to make 50–60 miles a day, changing animals often. I am anxious to see how this arrangement works out. It seems to me to be about the best way of getting over the ground in a hurry.

"The woods are full of geologists and men taking out concessions and the company is going to have to get busy if they want to get hold of any territory. All this province seems to be quite likely country, and may develop into an oil field in time. Such oil as I have seen so far is of asphalt base, and very poor in gasoline, and gasoline is what is most needed nowadays.

G.L. Harrington with mules north of San Pedro, Jujuy Province, Argentina, in 1921. R.H. Dott.

"The geologist Bennett, mentioned in the January 13 letter, was working for Standard Oil Company of California."

The 400-mile round trip to the Rio Grande took them approximately to latitude 36 degrees S, longitude 70 degrees W. From Buenos Aires, Dott wrote his father about it on February 24: "...I was gone down there about 3 weeks, getting back to San Rafael on the 14th; and leaving the same day for B.A. where I arrived the 15th. We put in long days, and covered a lot of ground, making 15 to 18 leagues a day, often. The country—or nearly all of it—is a pampa, flat as a floor. The southern point, near the Rio Grande, is somewhat mountainous.

"We went first to a big ranch [Sosneado] where some wells have been drilled. The ranch has some 370,000 hectares in it, and is owned by some English people. The manager is a Scotchman, who treated us very nicely.

"The ranch house is 100–150 miles from San Rafael, but it is fixed up in fine shape, with electric lights, fine baths, electric heaters, etc. all-in-all a white man's layout. Only a small percentage is under fence or cultivation, and all they raise is alfalfa. Cattle are the mainstay of the ranch. They run them in the mountains during the summer, then bring them out onto the pampa to graze in winter and also feed them alfalfa.

"From this ranch, we went to a town called Malargue or Cañada Colorada. This was a small place built up around another ranch house.

"The area is very dry, so that irrigation is necessary in order to grow anything. Also the wind blows very hard. On the road leading to Malargue we passed great sand dunes. Bevan, the Scotch ranch manager, said he had seen the wind blow a man off a horse; and Kirkpatrick said that on his other trip, they'd had an awful time to saddle up, on account of the wind. But even with wind and hot sun, which, reflecting on the sand burnt my face to a blister, I much prefer Mendoza to Bolivia and tropical Argentina.

"From Malargue, we went south to the Rio Grande, thence west, along the north bank to a latigo, or cable over the river, by means of which we got across. From the place where we hit the river, to the latigo, we went over a high mountain to see some fossils, which didn't materialize. On the way over, we went up about 9,000 feet and into at least a 90-mile gale. I was really afraid for a few minutes that I, mule and all were going over the edge, and it was a long way down.

"After crossing the river, we went south, saw some rafaelite deposits—interesting, but not very important to us.

["Rafaelite" belongs to the group of solid hydrocarbons called asphaltites, and is named from the town San Rafael. It is commonly mistaken for coal. During W.W.I it was mined for the metal vanadium recovered from the ash. The deposits are in veins and probably were formed by natural distillation of oil out of shale by igneous intrusions, with loss of lighter constituents.]

"The people down in that country seem to be as poor as can be. They live in little straw

Peons load up the mules for the trail in Jujuy Province, Argentina. R.H. Dott.

Geologists put their Ford to use hauling a rancher's Maxwell up the hill near Comodoro Rivadavia, Argentina, in 1921. R.H. Dott.

huts, raise goats, sheep or mules, perhaps cattle. Some of them are no doubt quite rich, despite their appearance. But they are awfully backward. They like potatoes, onions, etc., but rather than go to the trouble of growing them, they go without potatoes, and one of them had some onions he had packed 5 days on mule back from Chile. Most of them are Chilenos. They raise a little alfalfa, but never cut it, so the cattle go pretty hungry toward spring.

"I saw some interesting country but no likely territory. I wanted to get back in a hurry, so as to make the Monday train—they run a tri-weekly (try every week) schedule. On the trip from Malargue to San Rafael, we traveled till 9:00 and 10:00 p.m. twice. The third night out, we got to within a mile of our goal for that night, at 10:00, but my horse was all in, and we had left five along the road.

"Next a.m. we got to the bridge we were aiming for, and bought some alfalfa for the animals. Kirkpatrick left the horses there with our peon, who was to recover the stragglers and then take the string to his place. We went on to San Rafael on a string of mules which we had gotten in Malargue, where Kirkpatrick had kept them in pasture.

"We could have made San Rafael in a day with an early enough start, but didn't leave the bridge until 10:30, so stopped about 7 leagues short of town. Next a.m., we got to a place out about 4 leagues and there phoned for a Ford. I got in about 11:30, cleaned up, tended to some business, ate, then bought my ticket, checked my baggage and hopped on the train about two minutes before she sailed."

Dott found Mertie, Frasch and King in Buenos Aires; Mertie was finishing the report on his work in Neuquen, and anxious to get started back to the States. Dott helped him finish his map, and Mertie left February 23 for Valparaiso, Chile, to catch his boat to New York. Continuing the February 24 letter, Dott wrote "...he sure was tickled but I was anything but that, for it made me want to go, too.

"A. G. Maddren, an ex-U.S.G.S. man, arrived Tuesday, February 22 and Stebinger, with one other man will be here next week, and another man about a week later. The company seems to be going after it in earnest. I'll be anxious to know where I am to go next...

"I wish you would come down and see me. I'd guarantee you a good time, but don't know how you'd like the horse and mule-back riding. But I do need a good cook, and think I could pay 100 peso ($30.00) per month and expenses, so you'd better come along.

"I am now working on a report of the area I covered. I haven't much of a report to make, for I went over it so fast, but guess I can give them something."

Dott's diary entry for February 28, 1921, confirms his observation in the February 24, about the company going after it in earnest. "Stebinger and party got in: Stebinger and wife, Eberly, Eskerson and wife. Eberly and Eskerson are land men."

A Buenos Aires office had been established when Dott first arrived in December, 1920. As a front to conceal the Jersey Standard connection or at least minimize it, the listing was "Eberly and Stebinger, Geologists and Engineers." Dott noted in his diary: "I doubt if anyone is fooled."

Diary, March 7: "Harringtons got in in the evening from Valpo." G. L. Harrington, formerly U.S.G.S., had been in Bolivia with the E. W. Shaw party working on oil exploration for the Braden Copper company. He was now back, working for "Eberly and Stebinger" and, in fact, stayed until retirement.

Diary, March 23. "Frasch, Maddren, and Birkett got away." [Probably for the north.]

Harrington and Dott were assigned to Comodoro Rivadavia, 1,000 miles south of Buenos Aires on the coast, where the Argentine navy had accidently discovered an oil field while drilling for water to serve a naval station. On March 24, 1921 the geologists left Buenos Aires by train for Rio Colorado, with a helper named Christy who was well acquainted around Comodoro Rivadavia, and a commercial attaché of the American consul, named Brady, who hooked a ride to Comodoro Rivadavia. At Rio Colorado they found waiting for them an unusual vehicle consisting of a Ford touring-car body (for passenger comfort, including a top and side curtains), mounted on a pick-up truck chassis (for power); in the space behind the body was mounted a wooden box for grub and cooking utensils.

Leaving Rio Colorado on March 26, they arrived in Comodoro Rivadavia five days later. They had a fairly good road along the coast, with many gates to open en route. There were many opportunities to shoot a game bird, similar in size and taste to the pheasant, which they got cooked at overnight stopping places. The bird was hunted commercially and could be found on menus of Buenos Aires restaurants.

The second night out they stopped at Port Maddryn, a town and surrounding agricultural area that had been settled by Welshmen. The area around Comodoro Rivadavia contains many Boers—mostly descendants, but a few original migrants—who had come from South Africa after the British takeover in 1899. Sheep raising and wool are the principal industry.

The geologists had a two-fold assignment—Dott to make a subsurface study of the oil field from well and production records; and Harrington to examine the environs of the oil field, looking for areas that might offer additional possibilities. Their arrival, and Harrington's departure, were duly recorded in the local press:

De Buenos Aires los señores: Major Jorge Brady, Ing. Guillermo Christi, Ing. R. H. Dott, e ingeniero G. L. Harrington.

—Para Deseado, los señores Ing. G. Christi e Ing. G. L. Harrington.

The first few days of April 1921 were spent getting organized, and meeting various officials of the navy and of the few private companies who own concessions and a few wells outside the boundary of the government reserve. Dott's diary entry for April 3 states: "Met Sullivan, of the Anglo-Persian outfit..."—additional testimony that Jersey Standard was not unique in its interest in overseas exploration.

Several entries in Dott's diary refer to "made candy in eve." His office and living quarters were a large bedroom in the Hotel Colón, where he also took his meals. Some of the camp gear was stored here, including a "canned heat" stove which he used to heat water for tea

At Camp Rio Sauce, Jujuy Province, Argentina, in 1921, "Cameron," an American, the interpreter and the cook, fixes breakfast in an "al fresco" camp. R.H. Dott.

and on occasion to make candy—mainly fudge. Such dry staples as sugar, salt, tea, and coffee, were carried in cotton bags for compactness and ease of packing. One candy entry in the diary reads "tried to make fudge out of salt—no luck." He had mistaken the salt bag for sugar.

Dott occasionally went to the field with Harrington and Christy and ran the alidade for Harrington. He taught Christy how to do it and, with more well data becoming available, devoted his time to office work.

Winter was coming on, and the May 11 entry records: "Cold as blazes. All stayed in." Diary entry, May 31: "...Christy left for B.A." There are few entries for the next two weeks. June 18: "Started out, but turned back acc't of snow."

On June 26: "[Steamship] *Metrie* in at 10:00 a.m. Left C.R. to go aboard at 4:00 p.m. Rough sea—got a bit wet. Were taken to ship in a barge. Have a nice cabin...Good eats." June 27: "Still in C.R.; passengers and ship's papers ashore. Very heavy wind, so impossible to send launch ashore...Not a bad boat." June 28. "Left about 3:00 p.m." June 30: "In Maddryn till 10:00 p.m. Had a bad eye—from cold or something. Took tea at Capt. Ross's table." July 4: "Landed B.A. at 10:00 a.m. Had to take baggage thru customs. [Comodoro Rivadavia is a duty-free port]...Had a hard job but finally got fixed up."

Dott spent the next seven weeks in Buenos Aires preparing his report and maps on the Comodoro Rivadavia oil field, for transmittal to New York. Along with work, he enjoyed the sociability and hospitality of the members of the growing staff, and their families. Additionally, he sampled the culture of the metropolis: horse racing, art galleries, movies, and especially opera, at the famous Teatro Colón in season, and lesser theaters at other times. It became a life-long addiction.

In addition to more geologists, negotiators, landmen, and production advisors from Carter Oil Company and other Jersey affiliates came to help advance the new (for Jersey Standard) venture. Major Tom Armstrong, as a negotiator, had much to do with the company's entry into both Bolivia and Argentina. Jack Conry was production superintendent for Carter, and Eberly had been a land man for the same company.

Dott's diary records that on July 14 Eskeson got back from a trip to Mendoza and Major Armstrong came also; and on July 22: "Stebinger, Conry and Christy off to Comodoro." Finally, on August 24: "Got two N.Y. copies of Comodoro report into Conry's hands at 2:30."

August 27: "Left B.A. at 9:30 for Salta...", where he met King August 30. They went on to San Pedro, and Calilegua, where they arranged with the manager for Leach brothers for the loan of animals, and hired peons and guides. King returned to Salta, and on September 7 Dott began an investigation of an area in the Province of Juyjuy that was known to contain seepages and several asphalt mines. With him was an American named Cameron, who had been hired in Buenos Aires as interpreter and cook. Before leaving San Pedro he learned that Frasch and Maddren were in Oran.

The unexplored areas of South America might be barren, but the cities included sidewalks and well-tended gardens. Gilbert P. Moore, Jersey Standard geologist (right) posed with fellow-passenger Micou in Montevideo, Uruguay, as they headed back from the Argentine June 25, 1922. R.H. Dott.

September 8: "Went to Garrapatal, ['place of the ticks'] where they are mining asphalt. Met Don Carlos, administrator, and showed me around and later gave me a guide to another deposit in Rio de la Brea..." That night in camp they discovered the place was aptly named.

The next week was spent following lumber trails that criss-crossed the grain of the regional structure, seeking reversals of dip that would signify an anticlinal axis. On September 14, 1921 they crossed the divide at 5,000 feet and saw snow-capped mountains to the west; they went on to Juyjuy. The next day: "Went toward San Pedro, but got only to where the river crosses the anticline, where we camped...One of our peons had spent all the supply money for booze, and wasn't in any too good condition. Saw no outcrops until 4:30, when I found the west dips..."

Dott was back in San Pedro on September 22. "Animals in bad shape." Harrington got in on the 24th, having finished his assignments. He reported Smith, another American

Robert H. Dott and J.E. Hawley take a break in the pleasant park-like atmosphere at La Plata, Argentina, February 19, 1921. Straw hats and canes were proper for the more civilized areas. R.H. Dott.

hired in Buenos Aires, to have been as bad as Cameron; they paid them off and sent them back to Buenos Aires on the 27th. Harrington took over part of Dott's area and they were back in the field. On October 10 Dott was back in the Calilegua area: "Met two men on road, one carrying a small coffin on horse back. Other made me take a drink of some kind of high-power booze [probably *aguadiente*—"tooth water"]. Hot day. One of the peons sick, and mule's back sore, so camped at 2:00 p.m." October 11: "To Rio Zora, and up about 3 leagues...Pack mules got stuck in quick-sand, and my bedroll got wet." On the 12th: "...Saw Harrington's chocolate tracks [paper wrappers] along road. Got into San Pedro at 3:00." October 14: "King and Premoli [land men] arrived at 11:15..., and Harrington at 3:00. King and Premoli left...for Juyjuy. They're going to make some fillings [for concessions]." The company was not wasting any time!

Harrington, King and Dott got to Buenos Aires on October 20; there they met a new arrival—a production man named Waldon. Frasch was back, and Maddren returned two weeks later. On November 29: "Man named Morris—an engineer—is here from La Paz [Bolivia]."

Dott was back in Salta December 12, and on to el Quemado, Juyjuy Province, to fill in details time had not permitted him to get in October. He stayed with Leach Brothers' manager, a man named Kelly, who also loaned him a horse. Dott noted December 18: "...Fine dinner. Kelly sure is a fine host." The rainy season had set in, streams were up, the bush was wet much of the time and uncomfortably hot, and bugs terrible. In a letter to his father written Christmas Day 1921, Dott commented: "...[this] is really no place for a geologist at this time of year." Diary entry, December 23: "Went to Rio Negro. Spent day in beating through the bush after outcrops. Much grief but very little results. Rio Negro so high, impossible to pass..." December 24: "To Rio Negro again. Tried to cross and almost got drowned trying to swim it. current too swift..." December 25: "All alone. It is a strange Christmas."

On January 7: "...Peon killed fish with machete and I had it for supper, at hotel [San Pedro]."

Dott returned to Buenos Aires January 12, 1922, and learned that Mr. Bowen, the New York boss, was in Neuquen. He also learned that his own next assignment probably would be there rather than in Bolivia. He noted in his diary: "I am lucky newcomers have to go to Bolivia and northern Argentina in summer. Poor devils!"

Back in Buenos Aires, Dott found that Bela Hubbard and Gilbert Moore had recently arrived from Peru, having crossed the continent via the Amazon and its tributaries. In late January, Moore left for Comodoro Rivadavia. Hubbard left later for Plaza Huincul, Neuquen, not far from Dott's area. The Government had a small oil field near Plaza Huincul, and the Jersey company was starting to drill in that area.

Other newcomers early in 1922 were Mr. and Mrs. E. F. Burchard, U.S.G.S., two men named Schultz and Brammer, who seemed to be working with Armstrong and

Stanley C. Herold, geologist for Sinclair Oil Co., in 1922, poses in field regalia. R.H. Dott.

Dott left Buenos Aires February 23, 1922, for Zapala, Neuquen Territory, the end of the line, to make a plane-table map of a large anticline that Stebinger and Bowen had seen on their trip in January. A couple of Anglo-Persian geologists on the train were headed for Challaco. Dott's destination was an estancia two miles north of the railroad, owned by a Frenchman named Plantez; it was located on the Neuquen River at the foot of an escarpment capped with sandstone, that rose 1,500-2,000 feet above the river.

Dott found several willow poles, 10 feet long and 1½ to 2 inches in diameter near the water, and carried them to the top of the escarpment. He selected sites on commanding promotories and erected the poles, with white cloth attached, in rock cairnes to withstand the wind. Two of these could be used for triangulation. He then ran an approximately half-mile base line on the valley floor. Finally, with the aid of a peon, a stadia rod, plane table and alidade, assumed location and elevation above sea level, he was prepared to establish the relative location and elevation of any point desired.

During most of his stay Dott camped near the ranch house and took his meals there. The work went well. After fighting through brush and finding only small scraps of outcrop, this wide-open country with everything well exposed seemed like a geologist's heaven. There were structural complications that proved very puzzling and challenging.

Geologist Bela Hubbard working on the planetable in Neuguen Territory, Argentina, 1922. R.H. Dott.

On Train
2/23/22

Diary: "Left at 6:40 for Zapala. Some Anglo-Persian people are going to Challaco. Robinson and wife, Holmes (an Englishman) and Wolverton, a driller."

Zapala, Neuquen
2/28/22
RHD to RMD.

"I came through Challaco at 8:00 a.m. Suturdaz, and saw our rigs about 9 miles to the north. They have hit a strong flow of gas in two of the wells, but haven't reached the oil sand yet. [Sand that produces in Y.P.F., or government field at Plaza Huincul]. Mertie, Stebinger, Bowen and Maddren had made reconnaissances of this region earlier. I'm working in valley of Neuquen River, near P.O. and telegraph station called Paso de las Indios."

Camp, near Paso de los Indios
RHD to RMD
3/2/22

"There's a government surveyor here now...He is making a topographic map of this area; 2,000 sq mi to map in six months. His equipment is a compass and a watch, to determine distance. That's the way most of the Argentine maps are made."

Juan Parada, one of the peons, carries the poles for a triangulation station to be placed on top of El Mangrullo escarpment, March 7, 1922. Parada holds a pichi (armadillo), which the Indians roasted in its shell. R.H. Dott.

Geologists help set up the campsite for drilling Jersey Standard's first well (Agrio) in Neuquen Territory, Argentina, in 1922. R.H. Dott.

Zapala
5/17/22
    Diary: "Found Frasch, Hubbard, Lane (Frasch's instrument man) and Knight—production man in charge of Agrio [I believe that was the name of the concession the company was drilling]."

5/18/22
    Diary: "Decided to return to Paso de los Indios to take a reconnaissance trip to the north of my other work, north of river."

Camp
5/20/22
    Diary: "Wire from Stebinger says to put in 2 weeks at Rosales before going to B.A."

Rosales
5/27/22
    Diary: "Arrived at Rosales at 4:00 p.m."

Conditions were often extreme and methods crude and labor-intensive, but effective. Here, Dott observed workmen cutting and squaring large cedar timbers, in the Province of Sulta.

6/1/22
   Diary: "Finished [Rosales job] at noon, too late to start for well, so will go tomorrow."

Agrio
6/2/22
   Diary: "Got to Agrio camp about noon; found Harleman at work building camp buildings."

Ramon Castro
6/4/22
   Diary: "Has a good Chinese cook."

Oil had already been discovered in Argentina when Standard Oil sent its men into the area. The Mina los Bujitres on the north slope of the Cerro Alquitran, Mendoza Province, near El Sosneado, had been developed between 1908–11. Barrels are lined up to be filled and transported by cart to San Rafael. R.H. Dott.

6/4/22
    Diary: "After turkey dinner went to Ramon Castro in truck. Letter from Stebinger; looks like my transfer is going through."

On train
6/5/22
    Diary: "Got to Zapala at 11:00. All the rig material is in Zapala."

6/24/22
    Diary: "Left B.A. for New York; arrive 7/11/22."

Thus Dott ended his sojourn in South America. It was fruitful for him and for Standard. As Dott remarked to his father, the "woods [had become] full of geologists." Indeed, the activity in South America began a new era of exploration as companies sent geologists to every corner. Each was, like Dott and his companions, rugged and resourceful, adventurer as well as scientist.

Working for the Survey

Every postgraduate geologic student had one eye on employment by the U.S. Geological Survey in Washington in the early years of the century, even after World War I. J. Brian Eby was one of them. To qualify, each person had to pass the geologic aide examination before they submitted an application. The director of the U.S.G.S. in 1920 was George Otis Smith, a Johns Hopkins doctor of philosophy in geology.

Although Eby's exam score was low, he managed to get in because of his ability to write and his former training as a newspaper reporter. He was sent to Virginia. "After I arrived at Big Stone Gap early in the summer and checked into the local hotel, [Chester K.] Wentworth casually asked me if I ever mapped an outcrop by plane table. Truthfully, I said no. Taking me to a second floor hotel window he pointed to a large fenced field that was perfectly flat, except for a huge sandstone rock near the center. The line where the soil of the field meets the edge of the rock is a geologic outcrop. Tomorrow morning, he said, I would map it and if it took more than ten minutes I would crack up that boulder with a sledge hammer for a week. Recalling my plane-table and transit experience in the ROTC and the Army, my outcrop lesson was done in nothing flat."

After the first few weeks, Wentworth and Eby paired up to work alone or with a native guide. Plane-table work was impractical in the area, so they used a Brunton compass, a chisel-edge hammer and aneroid barometer. With that they carried notebook cases, a canteen and lunch. If coal sampling, they handled a 30-pound sampling outfit. They used a Ford car when and where roads existed.

Eby managed well with the native mountain people and even found the mountain moonshiners hospitable—to a point. "Mr. Horsley, the Big Stone Gap postmaster, suggested that on my mountain treks it would be prudent to hire a 'reliable' guide for a few trips to get myself acquainted with the mountain brethren," Eby wrote. "This I did. Newspaper accounts told me that one county, state or national lawman had been murdered every month for the past four months in the county. My khaki clothes, big hat and leather goods made me look 'powerfully suspicious.' How the news got through the mountains I'll never know, but when I wandered into one still in full bloom I was not only welcomed but offered a mug of 'mountain dew.' I settled for undiluted and unpolluted spring water. I kept their friendship but sure lost their esteem."

Moonshiners were not the only danger in the mountains. Eby was walking along the edge of the river in the five-mile Guest River Gorge in central Wise County, Virginia, on a ledge about 15 feet above the water. "Suddenly I stepped into a concealed crevice in the sandstone floor and I fell about eight feet until the crevice narrowed to about 20 inches, pinning me between the walls, wedged in place by the knapsack strapped to my back. I had no cuts or bruises, so I used the principle of the inchworm, fists and elbows for top leverage—and my hobnailed boots as bottom leverage. Once I reached the top, I spotted my wide-brimmed Stetson still carefully draped over the point of my entry. My only loss was

While many geologists were exploring foreign countries, others were active in the more remote areas of the United States. Sidney Paige examines an oil seep at the lower end of Water Pocket Fold, Garfield County, Utah, October 5, 1921. E.C. La Rue/U.S. Geological Survey.

Geologists pose on top of Wasatch Plateau, Ephraim road, Sanpete County, Utah, June 20, 1922. The side curtains of the vehicle rolled down in rain or snow. Water was carried in the canvas bag, with the tool kit on the running board. J. Brian Eby is kneeling in the snowbank. E.M. Spieker/U.S. Geological Survey.

my chisel-edged hammer and professional dignity. Recovering the latter, I thanked the good Lord for the pardon and went on my way."

Into a New Century

Activity in the Wyoming fields was going strong in 1920, and the northern part of the United States still thought of itself as western frontier.

Charlie Hares, then with the Ohio Oil Co., found himself busier than ever after the war. When he could no longer handle the number of locations that needed examining, a second man was hired—Dr. Wilson B. Emery of the U.S. Geological Survey. They set up a specific procedure. Hares' ranged through the several states, spotting likely areas. Emery followed for pinpoint testing and mapping. After surface scanning, they made test drills and took cores as deep as 200 feet to correlate subsurface formations with visible outcrops. In this way, they were able to plot trends or folds and stake wildcat locations. They also sought drillers' logs; but Ohio Oil cared little for such records. If a foreman bothered with them, he usually prepared them after drilling and guessed at many of the entries. To gain reliable deep sampling, Hares and Emery kept close watch on the data themselves. Gradually and—as Hares thought—with some reluctance on the part of the company, he and his cohorts were able to pull The Ohio into the new century with geological procedures on a sound scientific method.

Ed Owen maps a surface structure in eastern Kansas in the 1920s.
E.W. Owen/Mirva Owen.

Escape from Montana

Late in 1920, Ed Owen found himself in Winnett, Montana, where he wrote his former professor, W.A. "Doc" Tarr. "For the past six or seven weeks I have made Winnett my headquarters and have had some interesting problems to work on," he wrote on November 5. "I worked all of Brush Creek, the west end of Cat Creek, the Ingomar structure, Calf Creek structure (Garfield Co.) and a few others of less importance.

"[K.C.] Heald said he had [Glen] Ruby & Cave in the northeast part of the state. He said both were making good and believed Ruby would make an exceptionally good man as soon as he ironed out a few of the kinks...I certainly have been meeting some big men in the oil line and all kinds of geologists...I have also had to refuse four different parties who wanted me to do some consulting work for them...

"Well Cat Creek has five big producers and I expect three more in by the end of next week. One of these the Anaconda well will extend the field 2 miles west if it is a good one and indications point that way... P.S. We have a "doodle-bug" expert in town."

Young men who worked with the U.S.G.S. parties and state surveys could find life tough.

American roads were often as bad as those encountered by geologists in foreign countries. This bridge in Utah consisted of a series of logs thrown across the stream. A driver needed a sharp eye to make it across.
E.M. Spieker/U.S. Geological Survey.

In June 1923, Professor E.B. Branson of the University of Missouri took a reconnaissance trip west from Shiprock, New Mexico. With him were Ray R. Moody and Virgil B. Cole. "It was a very difficult country to drive across," Cole recalled. "Much of it was solid sandstone and windblown sand. We made it thru by following 'rock johnnies,' built by Navajos so that they could find their way home from the trading post. Upon leaving Goodridge, Utah, we had to find our way across a dim trail, covered by wind-blown sand and sand dunes. The Professor drove of course, and we pushed. A rear axle was snapped trying to get thru a small sand dune. We were 149 miles from the nearest railroad, no roads, no white people, and few Navajos. Fortunately for us, Wilson Cramner Co of Denver, was operating a core drill near Organ Rock. Ten days and a severe sandstorm later, Branson returned with a new axle. He was accompanied by A.I. Levorsen, on his first job with Gypsy Oil Company." V.B. Cole.

In 1921, J. Brian Eby left Virginia to head out west to survey Montana with a U.S.G.S. party. Winifred was one of the last of the old wild west towns, according to Eby. He and Frank Reeves checked into the only hotel in town to spend a few days restocking their supplies. "When checking into the hotel, a typical two-story wooden building, the clerk asked if I preferred the quiet side or the noisy side of the hotel," Eby wrote. "I chose the quiet side, facing the back yard enclosure. The reason for the choice was disclosed from 2 to 3:30 a.m. Friday and Saturday nights as roughriding cowboys rode up and down the main street yelling and shooting to celebrate their arrival in town for a weekend."

Reeves's crew consisted of two plane-table parties that were mapping geological formations of the Judith River basin, north of Winifred and south of the Missouri River. Reeves was party chief, M.N. Bramlette was the geologist, and Eby was the instrument man, plane-table operator, or geologic aide.

"It was my first trip west and we lived in small individual tents, usually pitched alongside a rancher's home from whom we bought our groceries or meals, or simple supplies," Eby wrote. "If we needed horses, we would rent them by the day, but largely we operated by Ford or Dodge automobiles."

The group hoped to complete their surveys before winter, but they had not counted on an early season. "Our escape from Montana was dramatic," Eby wrote. "We were to leave about September 15 to beat the winter snows, but we didn't quite do it. We were camped beside the homestead of the Hess ranch 20 miles north of Winifred the night before our scheduled departure when a snow storm pounded us for five consecutive days, the temperature dropping to 17 [degrees] below zero. Our canvas cots were so cold that I turned mine upside down, placed some pine cuttings on the ground, wrapped myself in blankets and crawled under the cot. I survived and two weeks later returned to Washington and Baltimore."

Tragedy struck this survey team in Yellowstone Park in 1924. J. Brian Eby (right) supports U.S.G.S. chief draftsman Clark, who died the day after the photo was taken. J.B. Eby/Gulf Publishing.

The Pack Rat Syndicate

In the spring of 1922, consulting geologist Ralph Arnold received a letter from 16 businessmen in Kalispell, on the western slope of the Rockies in Montana. The men were prominent members of the Chamber of Commerce, and they wanted to "check on a discovery of asphalt."

Arnold tried to beg off from the job. "I was morally certain there were no oil-bearing formations within miles of Kalispell," Arnold wrote. The group telegraphed me half my fees and said I had to come personally. So I soon found myself in Kalispell, ready to be shown the asphalt. For three days I was driven all over the country adjacent to Kalispell, and never saw a single sign of oil or oil bearing formation. In the afternoon of the third day I was taken to a deep, narrow, box canyon near the northeast corner of Flathead Lake. The seepage or deposit was said to be in this canyon, and we were to walk to it. The sides of the canyon were of cavernous limestone and nearly vertical. Some of the caves were of considerable size. Finally the deposit was pointed out to me. The material looked like gray raw rubber, and occurred in a vertical crack extending up to the floor of a large cave 20 or 30 feet above the bottom of the canyon...The cave was full of pack rats' nests. I had seen similar conditions in a cave in Solidad Canyon, east of Saugus, Los Angeles County, California. There, also, there was an 'asphalt' deposit.

"We climbed up into the limestone cave and I pointed out to my associates that the floor of the cave was covered with the excrement of the rats and that it sloped gently toward the crack in which the 'asphalt' was formed. The explanation for the occurrence of the asphalt was crystal clear—the excrement of the rats had been collecting for years in the crack, and thru the agencies of gravitation and water and air, the material had been converted into the raw rubber like substance which was oozing from the bottom of the crack."

It was a real blow to the would-be oil investors. Not only were their egos bruised, but they had already arranged for an open meeting of the citizens of Kalispell that night. They had expected Arnold to disclose finding of asphalt and the opening of a great oil field.

"I told them I would handle the meeting but that they would have to take the consequences of a terrific ribbing from their fellow citizens." That night, the hall was packed to the doors, "it being a free show," Arnold wrote. "I was introduced in glowing terms and soon launched into a general statement about the oil business. Then I stopped suddenly and after a moment's pause, began telling the crowd about a little animal known to them all as the pack rat. The crowd thought I had gone nuts. I explained that the pack rat was a cleanly little animal that always went to the same place near his nest when he had any duties to perform. Then I told them how the excrement had gotten into the crack from the floor and had been converted into 'asphalt.' By this time the meeting had simply gone to pieces. The promoters of the great oil venture had been thoroughly, good naturedly teased, and the crowd melted away.

"I was called into the directors' room of the Chamber, was given the balance of my fee

Dr. M.G. Mahl, University of Missouri, secures part of a fossilized Phytosaur in the Painted Desert in Arizona. Mahl was one of the founders of AAPG. With him is Thomas M. Nelson, North Texas State University faculty, and Chester D. Whorton, consulting geologist. J.M. Clark.

As well-site geology became more popular, many geologists found themselves working closely with the crews on the rigs. This Humble rig in Haynesville, Louisiana, in 1924, was built on a pine-log base. Humble Way.

and expenses and was ready to go to my train. I remained long enough to thank the syndicate for their sportsmanship and to tell them to raise another fund, about $3,000, the size of the one they used to learn about pack rats, and I would try to get them a real oil field. Half of the original men took up my suggestion—half of them dropped out, saying they had had enough of the oil business.

"A few days later Charles Emmons, geologist for the Kalispell group, advised me that he had found what he thought was an oil structure north of Shelby, Montana, and asked me to come up and check it with him. I...sent my office associate, Wayne Loel, to look over the prospective structure with Emmons. In about two weeks Loel came back to Los Angeles with a subsurface map of a beautiful oil dome which he and Emmons worked out in the field. I immediately approved the structure and leasing began on it at once. The Kalispell group used their $3,000 fund to lease 6,000 acres across the top of the dome on terms of 50 cents per acre for a 1/8th royalty lease, with option to purchase the land for $10 an acre.

"In the meantime a geologist...brought in a discovery well on the west flank of the dome. This started a wild oil boom in northwestern Montana. People came in from all over the country trying to buy leases. The Kalispell crowd ([Kalispell Kevin Oil Company]) opened an office on the structure, and subleased their land for as high as a 1/4th royalty, plus a bonus of $200 per acre. They never drilled a well. The first year they cleaned up $1,500,000."

Arnold had a postscript to the story. "As an aftermath of the wonderful success of the Kalispell Kevin Oil Company, the men of the original 'pack-rat' syndicate who had failed to subscribe to the second syndicate, sued the group who had put up the second money, claiming they should have had a portion of the riches. They lost their suit.

"You may look down on the lowly pack rat, but remember this," Arnold added. "If it had not been for the pack rats, there would have been no 'Kalispell asphalt' and consequently no Kalispell group to open up the Kevin-Sunburst oil field."

Too Much Ransom

Conditions appeared to have improved in Mexico by the early 1920s. Joseph Singewald felt the area was reasonably safe; and during the summers of 1921 and 1922, he worked with the Transcontinental Oil Company, a subsidiary of Standard Oil (New Jersey), on geological parties from Tampico to Coatzacoalcos on the Isthmus of Tehuantepec. "These parties were under the general direction of Raymond Leibensperger, a former student of Singewald's," according to J. Brian Eby. Chief of Singewald's party in 1922 was William A. Baker, also a Johns Hopkins University man.

They were working in a remote region of the state of Tabasco when Singewald was seized by Valentino Fausto, a guerrilla leader. Fausto proposed to hold Singewald for 5,000 pesos ransom. Singewald assured his captor that no one would pay 5,000 pesos for a simple

In 1923, companies were hiring geologists and establishing entire geological departments. Amerada's geological department had grown to more than twenty—including three women. Claude Williams.

The handgun was standard equipment when Willard Classen explored the Ecuadorian jungles in 1923. A geologist never knew what he might discover around the next bend or bush in the more remote areas. W.J. Classen.

university professor—that was the going rate for an oil man. Fausto agreed to reduce the ransom to 2,000 pesos. "Singewald paid the ransom out of his pocket, collected from Baker who, in turn, billed his company. The company treasurer in New York was hard to convince that one of their consultant's life had been at stake, but eventually paid."

Some were not as enthusiastic about prospects in South America. Outsiders often had difficulty judging the worth of geology because geologists appeared to be erratic. In February 1922, Ralph Arnold, Barnabas Bryan and George A. Macready delivered a paper at the New York meeting of the American Institute of Mining and Metallurgical Engineers. "As regards South America...many of us are gazing into a mirage," they wrote. "Ten cents spent in increasing the percentage recovery of oil from known American fields will be more productive of profits than a dollar spent in the tropics...Shale oils produced in the United States can be laid down in the great markets from Chicago and St. Louis westward, at less cost per barrel than oil produced from sand-storage fields in the tropics." Before the paper was set in type, on December 16, 1922, Shell struck a well in Venezuela that blew wild at an estimated 100,000 barrels a day. A subsequent well blew out at 10,000 barrels per day for nine days, until it sanded up. The La Rosa field was a major find, and the dire predictions were quietly buried.

Others—non-geologists—were often luckier than the geologists. In 1923, Rolf Engleman, then working for Metropolitan Petroleum in Mexico, visited Venezuela and was asked to return via Cuba to look over a structure south of Havana. There he met Clarence Owen, an oil promoter, who introduced him to his "geologist," a man by the name of Boulton. Boulton was a Canadian, "apparently the black sheep in the Boulton family who owned the Red D Steamship Line. He had spent his youth in Central America. He was there as a young buck at a time when all young bucks had to fight duels, and he fought several." By this time, he was 70; but he was not to be outdone by age. This "charming little rooster of a man" had a young wife from Spain and two babies. Like several others, he had earned his right to be called a geologist because he happened to locate a well—in his case, the first well in the Bacuranao field, east of Havana.

Blowout in Turner Valley

In 1923, Alex McQueen, who had been with Ted Link at Norman Wells, was wellsite geologist on Royalite (Imperial) No. 4 in Canada's Turner Valley. The well had been drilling off and on for two years, and the word was out to abandon all efforts. As they were about to give up at 3,500[?] feet, the well blew in with 30 million cubic feet of gas per day.

"We put a gate valve on the well to shut off the flow of gas, and then watched while the pressure on the gauge kept building higher and higher," McQueen recalled. "Finally we figured we'd better get the heck out of there." The crew watched from a safe distance as the force of the gas pushed the drill pipe out of the hole. "The pipe just started walking out of the hole until there was 60 feet of six- and eight-inch pipe standing straight up in the air," he said. The pipe settled back into the hole as the escaping gas relieved the pressure. But a spark ignited the gas, and the roaring ball of flame was visible 25 miles away at Calgary. It was several weeks before the fire was extinguished and the well tamed. It was the largest gas flow in Canada and opened even more geological efforts in the north.

One Dead Horse

Like geologists in other parts of the country, those who made field trips into Canada often found themselves up against unforeseen troubles—not only from the harsh environment, but sometimes from employers. One Standard Oil Company of California geologist returned to camp one evening to find that his horse had died for no apparent reason. He had been up against accountants without field experience before, and he figured he had a battle on his hands. It was well known among geologists that dead mules frequently had to be justified on the expense statements with the hide portion of the mule which bore the brand.

The geologist turned in his report and entered the cost of "one dead horse." He waited for a battle, but nothing happened. The silence piqued his curiosity. Finally he asked his superior why nothing had been said. The superior smiled, and told him that years before he himself had been a helper on a summer field party with the U.S. Geological Survey. It was his job to look after the mules. One mule had been particularly good at slipping the hobbles, and they had had to build a special halter for her. Even then she would curl up in all sorts of peculiar positions trying to kick the halter off. One night, she had managed to tangle both a hind foot and a front foot in the halter around her neck. As she struggled to get loose, she had fallen over a small ledge and broken her neck. When the sun rose, he had found her, stone dead. At that time, a mule was valued at about $60, and the helper was paid only $90 a month. With great care, he detailed in writing on his government expense form the exact circumstances under which the mule had died. He even sent in a sketch. He submitted the form to Washington along with his other expenses. Three weeks later, he received a reply. The mule, they informed him, could not possibly have died under the circumstances he gave, and the account would not be paid. The helper had had to pay for the animal out of his own pocket. "I resolved that if I ever had the chance," the chief geologist said, "I would give somebody else a better break."

Milo Orr and his mount were a little mismatched. W.J. Classen.

Tom Galbraith, geologist for Standard Oil of California, poses in Ecuador in 1923. Galbraith, Orr and Classen made up the geological team. W.J. Classen.

When Sidney Powers traveled through southwest Arabia in 1923, he was accompanied by (left) "McGovern"—his assistant—and P. Laurare, a "French mineralogist." Also with him were three Arab soldiers (below, right) who acted as their escort. Sidney Powers/American Heritage Center, University of Wyoming.

Far Ends of the Earth

On the far side of the globe, four young American geologists landed at Batavia, Java, in March 1924. They had sailed on the American liner, President Hayes, from San Francisco to Singapore; then transferred to the inter-island steamship, Van der Wyck, for an overnight cruise to Batavia (Jakarta). Their trip had taken 35 days overall.

The leader of the group was Emerson M. Butterworth, then 31, a graduate of the University of California with eight years' experience in California and Alaska. With him were Richard N. Nelson, Milo Orr and Earl Wall.

They had been sent on a two-year assignment by Standard Oil of California to evaluate the sedimentary basins in the Netherlands East Indies and obtain exploration rights on prospective oil lands. According to Richard H. Hopper, it was the earliest effort of the Standard-Texaco group (Caltex, CPI) to obtain lands anywhere in the Eastern Hemisphere. The Butterworth team carried out geological exploration and mapping in north and central Sumatra, east Java, east Borneo and northwesternmost New Guinea.

"In the early years, geologists sent out from the head or district offices to do field work were completely out of touch for months at a time," Hopper wrote. "In many remote islands a monthly inter-island steamship was the only means of communication, and in the interior of Borneo there was not even that. Thus the principal means of contact from the geologist in the field to the head office was by letter, approximately once a month, to report on geological findings and to confirm that the explorer was still alive and well. Geological tent-camps were generally moved every two or three days, and it was exceedingly difficult for the head office to contact a geologist; in fact, it was frequently impossible for three or four months at a time. With respect to periods of duty in the field, the rule was six months of field work and one month of office work for unmarried geologists; for married geologists whose families lived in the headquarters city, the schedule was eased to three months and one month, respectively."

Supplies were handled like everything else—by hand and hard labor. "In those days, long lines of porters with loads on their heads threaded their way along jungle trails to keep the geological and seismograph parties supplied...Geologists in the remoter jungle areas were unable to eat fresh food for months at a time, living entirely on dried and canned food, hardtack and rice. Beri-beri, usually in a rather mild form, was common...Mosquitoes, leeches and ticks are among the smaller irritants of the animal kingdom, while the larger beasts which occasionally contest man's entry into Indonesia's tropical rain-forests include tigers, snakes, elephants and crocodiles."

It did not take them long to figure out how to protect themselves. "The simple precautions of keeping a few kerosene lamps burning at night in geological camps, and of not keeping live chickens, goats, dogs or monkeys in the camps, have proved nearly perfect in preventing attacks by tigers and panthers. Elephants, too, which sleep in the daytime and travel at night, will stay away from lights."

Navajos join Gypsy Oil Co. land man Amos Guthrie (extreme right) in March, 1924, to watch Gypsy drill No. 1 Tocito well on the Navajo reservation. V.B. Cole.

Harvey Bassler reclines on a balsa raft as he makes his way down the Rio Sarayaquilla in the Amazonian jungles of Peru in 1923. American Museum of Natural History.

The launch Beatriz was used by Harvey Bassler at the start of his 1924 expedition into the upper Amazon. A lighter is lashed to the side of the launch. American Museum of Natural History.

Hacking a Trail Westward

By the mid-1920s, Kessack White had migrated to Bolivia for Standard Oil of New Jersey. On July 20, 1924, he wrote Eugene Stebinger in Buenos Aires from Camp Sanandita, "We have located and mapped a well-defined dome west of Sanandita which we have named the Sanandita Dome...The well should be drilled as deep as possible...rigged for a depth of, at least, 4,000 feet." White outlined the trip to reach the dome, which included travel by cart road, cattle trail, pack animals up the river, with the last leg of the journey by foot.

Early in July, White set out to map the exposures on the Rios Blanco and Pescado along with "Murakozy." The man-pack expedition across the hills took four days. They slept in the open, and ate when they had a chance. While they mapped from one direction, another geologist surveyed toward them and "the capataz" and a dozen men hacked a trail from the Rio Pescado across the divide into the drainage of the Rio Bermejo.

After several days, it became apparent that it would take nearly a week to map the numerous ridges. They broke camp and regrouped to enter the country from the east. "We took what our mules could carry and stored the remaining in a strongly built hut," White wrote. It was difficult country, and White had to employ extra men to speed up the work.

A 12-kilometer gap between the Quebrada Colorada and the last section was "too far for man packing...as the base line must be carried by transit and several sections must be surveyed." But he knew it was especially important, "for if there is closure in Argentine it should indicate itself in this interval," he wrote.

His hunch paid off. White completed his reconnaissance in October and found himself headed for Buenos Aires. "I wrote you some time ago about my first location having been a rather good well [discovery well on the Bermejo River]," he wrote his mother. "It has now been drilled in and is a big well. I do not as yet know officially, but current report places it between 1,500 and 2,000 barrels. That puts this company on its feet. They are rushing work for my second location on another structure and have about completed the securing of concessions to drill my third. I am just at present in a very favorable position as having located the first commercial well for the company. The three previous wells drilled did not give enough oil to justify development. Two were dry and one got a little oil.

"The field season is now a thing of the past, and I am tired, for it was a hard year and we did very much work in the time we were out. I am on a very good train in a Pullman type chair car so that traveling is agreeable. The road bed is rough and it is running a bit fast so that writing is somewhat difficult. It is very gratifying to open up something that some day may mean a very large industry. But please do not speak of it outside of the house."

Shooting the Rapids

That same year, Joseph Singewald joined the Lincoln Ellsworth expedition to Brazil. Ellsworth served as plane-table operator, and Singewald was geologist-rodman. They traveled from Brazil, starting at Salaverri, eastward to the Huallaga River near the town of Pachiza, over the Andes to Peru.

After several months, Ellsworth became ill and returned to the United States. Singewald's brother Quentin joined him as plane-table man for the remaining months. They returned by floating down the Huallaga River on "balsa" rafts to Yurimaguas and thence by wood-burning river boats to the mouth of the Amazon.

Singewald returned to the upper Amazon of Peru in 1925, with Ernest Roschen as his assistant. They traversed via plane table the most awesome canyon of South America—the Pongo de Manseriches, whose whirlpools and rapids few white men had ever attempted (and few of those who did, survived). The river led upstream into land inhabited by the Jivaro Indians, who were notorious for their manufacture of shrunken heads.

The only geologist previously to enter the country was Harvey Bassler who, unlike Singewald, had been accompanied by an armed detachment of the Peruvian Army. In the Pongo, one of Singewald's canoes was lost in a whirlpool and another barely escaped a second whirlpool.

Not all the obstacles were natural, however. R.A. Liddle and D.P. Olcott had been given carte blanche to travel and compile a report on Venezuela and Trinidad for Standard Oil Company of Venezuela. It was rumored that they were thrown in jail in Port of Spain, Trinidad, and later escaped from the island in a row boat.

The Big International Argument

In 1925, an International Geological Party was put together to survey the new concession granted by the government of Iraq to the Turkish Petroleum Company. Ed Owen called it "one of the strangest organizations in the entire history of petroleum geology." The concession itself was makeshift with what several described as "onerous and impractical provisions."

The members of the party were representatives of their individual companies, except for three American geologists who were temporary consultants for the American group. The principals were still at odds over many issues; the geologists were not dedicated to a common purpose and were not sure they could trust each other. In addition, it was going to be a trying journey that called for what F.E. Wellings called "old-time qualities—strong legs, stout hearts, good bumps of locality, and the keen sense of smell for oil."

Leader of the party was Hugo de Bockh [de Boeckh], geological advisor with the Anglo-Persian Oil Company. Professor de Bockh was Hungarian, and had served as

Hungary's Under-Secretary of State for Mines. He had been in Persia in 1923 and spent two winter seasons there surveying the country.

"Almost from the start it became evident that it wasn't going to be a happy party," wrote Arthur H. Noble. "De Bockh was autocratic and had preconceived ideas about geology and the prospects for oil. He insisted that he must have his own Company's geologists in charge of the parties that were going to map individual structures. They were much younger than those from the other groups, and had no experience outside Persia. Almost at once [B.K.N.] Wyllie left the party—a great loss as he had a deep knowledge of the geology of the Middle East. The party included 18 geologists, among them E.W. Shaw, who had led several difficult and dangerous expeditions in South America; A.C. Trowbridge, 'a man of considerable experience;' Shirley Mason, who had been with Shaw in Bolivia; C.W. Creek, a young Shell man, and T.F. Williamson, a British geologist. Standard Oil of New Jersey was handling the affairs for the American interests."

"We were briefed in New York that we would find things different than working for an American company, and not to worry about it," Shirley Mason recalled in later years, "but tell them when we got home and they would straighten everything up, financially at least." The Americans arrived in London, then took a boat-train, a channel steamer to Cherbourg, and a train to Marseilles, where they boarded a ship supposedly for Beirut.

"While we were in Alexandria and planning to take a day off to go up to Cairo we got word that plans were changed," Mason said. "The Jebel Druse were up and we could no longer get in through Damascus so we left at once by train to Jerusalem where, after waiting a day or two, a convoy was arranged. There were about five big Rolls Royce and Cadillac touring cars with big water tanks and our baggage strapped on the runningboards when we took off." F. Wakeling, a transport man, recalled that the fleet consisted of 28 touring cars and light trucks to haul the men and the army of servants.

"We had a good road across the Jordan to Amman, but from there on our men were trying to follow the trail that had been made with a plow through the desert, to guide the R.A.F. in the first World War," Mason noted. "We were crammed in the cars, so enjoyed the opportunity to stretch out on the ground at night when our drivers got some rest. We could see the lights of the armored trucks and flashes that might have been gun fire where the French were settling their argument with the Druse."

The group reached Baghdad the following evening but had to wait, because there was trouble in the area. Political conditions were almost the same as in 1920. The Kurdish chief, Sheikh Mahmoud, who had been captured by the British and imprisoned in the Andaman Islands, had been released. He had returned and promptly resumed the war.

When the geologists finally got out of Baghdad, they could go only a little distance north and were not allowed to separate. "A troop of Levy cavalry reported every morning and our geologists weren't allowed to work until a screen had been thrown out in front of them," Noble wrote. Mason recalled that he walked along with "about 16 men in single file, each with his hammer knocking on the outcrops. A little later we were split into groups of about a half-dozen, and waited until the troops went out and took their position on the other side of the long range of anticlinal hills that we were working. We worked with one eye on the rocks and the other on the soldiers, so that if they started running we could take off and beat them to the walled town, Tuz Khurmatu, where we were staying. Fortunately, things quieted down, and at last we could work as individual field parties with only a mounted squad of rifle-carrying policemen to guard us."

On their first restricted trip out, the group rode on what remained of the Berlin-Baghdad Railway to the north. They then loaded up a group of camels. "After spending half a day loading, the camels promptly threw the packs off," Mason recalled. "We gave it up and loaded on donkeys that trotted off across the desert with their panniers just clearing the ground. Their jolting little steps were mighty uncomfortable."

On later treks, particularly in Qaiyarah, model-Ts managed to make it across the desert in fair shape. "We occasionally wrecked one when the native driver took off and didn't notice the pits that hunters had dug in an effort to stalk gazelles," Mason said.

Mason and Creek made their first main camp in the compound of the oil refinery at Qaiyarah, "a tin-pot affair making a yellow, corrosive gasoline for the R.A.F. trucks." The only oil production in the country was from two shallow wells in the compound which were pumped by coolies with long wooden sweeps.

"Our relaxations were heavy drinking parties in the R.A.F. headquarters in Mosul, and hunting," Mason recalled. "On one pleasant hunt we put off to an island in the Tigris on pontoon boats left over from the first world war, and moved up the island in a line, with each European with a gun separated by three or four coolies to beat the vegetation with sticks. Our good shooting was for black partridge, but we shot at the jackals and big white hare, and for the sake of the coolies I would shoot the owls because they knew that they could have those."

The main nightly feature was discussion at the dinner table—usually of the geological aspects of the terrain covered that day. "Being the only layman of the party, it was all double-Dutch to me," T.F. Williamson wrote, "but numerous terms and expressions so constantly recurred that I suppose I subconsciously absorbed a certain amount. One of these discussions in particular got quite heated and to break the tension I was asked for my opinion, in the hope, I suppose, of raising a laugh. Now was my chance to air my knowledge, and I proceeded to use every geological word I could remember in whatever order they occurred to me, with the result that I soon had them in hysterics. They then realized what I had to put up with every night and the leader of the party decided that there would be no more geology at the dinner table. In five minutes they were hard at it again."

Williamson's group followed the Nairn Transport Company's Desert Route, 530 miles from Baghdad to Damascus. "The trip was considered by many to be an adventure; others regarded it as a nightmare of heat and dust in summer, and of cold and mud in winter.

Convoys caught in the heavy rains often remained stuck in the mud for a week."

Williamson's party was lucky. On the return trip, the convoy consisted of three Cadillac touring cars. "We left the Nairn Company's yard at 7:30 A.M., and as the Euphrates was still in flood we had to circle south of Lake Habbaniyah. In the soft ground car after car stuck in the sand and we soon became adept at climbing out to help in extricating them. By mid-day we reached Ramadi, where lunch was produced, and a powerful brew of tea was poured into enamel mugs; and then we were off on another lap of our journey.

"Toward sunset we halted for tea during that period of half light when desert driving is difficult, and resumed our way in the rapidly deepening darkness. Slowly boredom overcame the passengers, who began to doze fitfully on each other's shoulders. The air became colder and soon jackets and even overcoats were donned and buttoned up. An evening meal was consumed in the light of the car headlamps and a short halt was made at Rutba where the fort was then being built.

"On account of the Druse rebellion, our route took us northwest to Palmyra in Syria...The cold, cramps and fatigue of the night were forgotten and we were all alive with interest when we reached the rest house, later the Zenobia Hotel."

When the group moved on, they spotted patches of ploughed land, "villages nearer to each other, and the harsh white glare of the desert gave way to a softer, kinder light. Red tiled roofs afforded a welcome change of colour and the running water and luscious fruits we found in Homs finally convinced us that we were out of the desert. Our car had at least a dozen punctures, several petrol stoppages and a broken spring."

Trouble with the Troops

Because of de Bockh's personal style and rather dictatorial manner, trouble quickly developed among the professionals. Arthur Noble did not like the way the mapping was being done, nor did A.C. Trowbridge. When Noble and E.W. Shaw approached de Bockh, a "stormy interview" developed. De Bockh refused to change procedures. Shaw and Noble put their feelings in writing to management in London and suddenly found themselves banished by de Bockh to do reconnaissance west of the Euphrates. Since Shaw and Noble got along well together, they at least were in pleasant company.

When the groups reconvened in Mosul for a final check of the surveys, the disagreements grew. "We found that de Boeckh had arrived at certain conclusions about the nature of the rock that was assumed to be the oil reservoir," Noble wrote. "It was not so much a difference of theory, but rather in the application of the theory to the assessments of the value of certain structures.

"As disagreement was inevitable, we thought it right to let him know that we should oppose his views when the final report was made," Noble wrote. "To our surprise we both received letters terminating our services at a month's notice.

"The result was that a very senior member of the Board [Turkish Petroleum Company] flew out to see what the trouble was about and the affair was settled amicably." The reports, however, with all the disagreements, were all registered and noted. The entire expedition went down in geological history as one of the most disagreeable working relationships ever concocted.

No More Uncivilized Country

Fred A. Sutton went to Argentina in March 1925 for Jersey Standard. The railroad had been extended 90 miles since July 1920, when the terminus was Embarcación. "Am located here in Tartagal, which is the end of the RR, and about 1200 miles north of B.A., and a few miles south of the Bolivian border," he wrote his wife, Anne (later Anne Sutton Weeks), on May 4. "I shall leave tomorrow or the next day for Bolivia, and have a 300 mile mule trip ahead of me...This is the company's distribution point for all supplies, etc., going into Bolivia. The supplies are sent in large carts which are drawn by as many as 12 or 18 mules. Transportation here is quite a problem, now especially since it is the end of the rainy season and the roads are knee-deep in mud. Our supplies will go...a very small part [of the way] in the carts and then we shall have to resort to pack mules...I have a Chinese cook, who I fear I shall have to sit on a couple of times to let him know who is boss."

Twelve days later, he was in Bolivia. "Our trip up here from Tartagal was rather slow on account of its being the first trip," he wrote. "The mules were wild and some of the men had never ridden before, so we had to just crawl along...They have a dandy camp (Buena Vista), good food, electric lights, and a shower bath. The company doctor is here at present. He has just come from up country where he went to attend one of the company men, an American, who was accidentally shot in the head. He died, however, before the doctor arrived."

The weather was ideal—cool days and cold nights. "I find it necessary to wear my sweater all day, and a coat at night. There are no mosquitoes or other pests now, but I had a bad attack of fever when I first came."

When Sutton wrote on June 27, he was in Cuevo. "The reason I am here...is because all my peons preferred going to the Cuevo carnival, to working, and so I fired them all. The fiesta is about over now, so tomorrow or the next day I shall hire me a new gang & return to camp."

Sutton was back in Cuevo in July. "For some time past, Cuevo has been terrorized by a number of bad men who have been entering houses at night, attacking women and taking whatever they desired," he wrote. "The local authorities were unable to do anything due to the fact that the subordinates of the Chief of Police were the principal bandits. I did not pay much attention to these happenings until these same bad men began attacking my peons. Every time that I would send a peon to Cuevo, they would catch him and beat him up. Also, while we were in camp, they entered the houses of these same peons and attacked their women folk. After this had happened a couple of times, I became fed up so the day before yesterday, Gunning and I went to Cuevo and rounded up six of these bad men and brought them here to camp where we are holding them prisoners until the arrival of some

If it was not too sandy, it was too muddy. Geologists find themselves stuck on what they called a highway, west of Tucumcari, New Mexico, August 26, 1925.
E.W. Owen/Mirva Owen.

soldiers from Lagunillas. I wired the office in La Paz and also wired Mr. Armstrong in B.A. to back me up in this...Please don't worry about me because I am able to take care of myself, and I have these people respecting me, and these Cuatreros [horse thieves] are certainly afraid of me." The men in camp shared the guard over the prisoners until the soldiers could arrive. They were returned to the local jail from which one escaped. The others were sent to Lagunillas for trial and were charged with rape and robbery. "As a result of our action, the President of Bolivia has wired the local authorities to take the most energetic action possible against said criminals," he wrote Anne.

Conditions were some better, but not totally quiet. A road engineer, a contractor and their chauffeur were murdered in a payroll holdup although at the time they were not carrying payroll. "A woman just came into the door to see if we had any cotton because a man up the street just now split open his wife's head with an ax," Sutton wrote. "The Intendente (chief of Police) here made a law that no Cuñas (Indian girls) are allowed in the street after 9 p.m. Therefore when he wants a girl to sleep with he sends a policeman to watch, and when a Cuña steps out of her house, she is arrested and taken to his house, and her fine is to spend an hour or so in bed with the Intendente...That is Bolivia. Rotten and corrupt. There is probably not a more uncivilized country in the world."

By fall 1927, H.W. Thoms, Eugene Stebinger, George Harrington and Sutton were busy marking well locations. "Yesterday [we] went out and made a well location," Sutton wrote. "It's Thoms' location because he worked the structure. They got oil in the 1st well & are going to drill another to prove up the structure. If this structure proves productive, I guess they will then drill mine that lies to the south of here. I am sure they will get oil there & I'd like to have one well to my credit before leaving this country."

Ask and Ye Shall Receive—Sometimes

As finding oil grew more complex, geological education grew more important. The Colorado School of Mines established a department of geophysics in 1926, with the cooperation of several oil companies. Graduates in geophysics were required to have adequate training in geology, as well.

One of the educational strongholds in the Midwest was the University of Wisconsin at Madison, where professors had been setting forth geologic theory since 1848. In the mid-1920s, Wisconsin had developed a strong, if somewhat eccentric, staff. There was Warren J. Mead, engineer, mathematician and mechanical genius, who was known to have purchased a Hammond typewriter when he became enamored of the internal workings, even though he did not type. The machine never did type, for Mead promptly took it apart to see how it worked.

There was also A.N. Winchell, whose text on optical mineralogy drew together for the

(Preceding page)
The Oklahoma Geological Survey organized a field conference to the Arbuckle Mountains in 1925 or 1926. Only part of the group have been identified. Standing, left to right: Robert H. Dott, R.B. Dunlevy, ?, ?, Fannie C. Edson, C.W. Tomlinson, ?, ?, N.D. Potter, Frank C. Clark, Grover Potter, ?, M.G. Cheney. Kneeling: John Fitts, E.O. Markham, ?, Frank C. Greene, ?, ?, Charles N. Gould (director, OGS), ?, ?, G.A. Waring?, ?, Luther H. White, George S. Buchanan, C.L. Cooper (?). Front: Kent Kimball. Among the unidentified are John M. Giles, John L. Hewendobler, K.H. Schilling, A.H. Koschmann, E.M. Mitchell, B.B. Hunter, H.C. George, S. Weidman and Don Gould.

first time a vast statistical compilation of important geological data. But Winchell was timid—too timid to make requests for equipment that was sorely needed in the department. Each fall, C.K. Leith would portion out the dollars at a fall luncheon. As it happened, the department was then equipped with microscopes that were best described as "truly antiquated." In fact, they were so old that Bausch & Lomb had refused to take them in trade on new ones.

Despite such a state of affairs, Winchell refused to ask for the sum necessary to replace them. Instead, when the meeting came round and the requests were accepted, he asked for one microscope and nothing else, although the department needed other items just as badly. At that time, microscopes cost only $300.

Leith was closing out the requests when R.C. Emmons, then a young instructor in the department, took it upon himself to do something about the pitiful state of equipment. Emmons had earlier approached all members of the department to back him when the time came. Now he requested one item—a dozen microscopes. Others echoed his sentiments. Winchell was awe-struck, but he was even more surprised the next week when the microscopes were approved.

There was Fredrick Thwaites, an extreme introvert whose strong opinions often set him apart. Even as a teacher, his techniques left something to be desired, since he talked in a nasal monotone. He was the frequent butt of student skits at Geology Club banquets; but he had a tremendously dry sense of humor that made him a favorite of students. He was also an avid field-trip organizer. During spring recess, he would gather a group for a week-long trip to the region around Devil's Lake. They would stay at a lakeside camp, accompanied by a cook and a chaperone. Rules of conduct were strict, but esprit de corps was always high. Thwaites gave students strict instructions: "Do not jump fences or off of cliffs," he frequently told them. "Do not talk to the natives—they will only delay and confuse you." A coveted award was sometimes given for the longest beard grown during the week, as measured scientifically by telescopic alidade.

One of the graduates of the school was Emily (Micky) Hahn. She obtained her baccalaureate degree in the Department of Mining and Metallurgy, in 1926, as the first woman recipient of an engineering degree from the University of Wisconsin. When Micky inquired about courses in the mining department which might help her in a geology degree, Chairman E.R. Shorey informed her that under no circumstances would his department accept a woman as a major student. That set Micky off. She immediately decided that she would change that attitude and went over Shorey's head to the college dean to achieve her goal. (Hahn would become a mining engineer, overseas geologist and author of 27 books.)

Many of the students who graduated from the university found employment with either the U.S. Geological Survey in Washington or with a nearby state survey. If they were lucky, they were among the chosen ones who also found jobs with the surveys during the summer recess.

Virgil Cole (back) and Roy Hall (right) pose with the "gals from the San Simon" Ranch at the original entrance to Carlsbad Cavern, February 1925. V.B. Cole.

From 1926 to 1928, L. Kenneth Lancaster worked in the mineral lands division of the Geological Survey, as a student employee during the summer. The assignment was a continuing project to map the Keweenawan flows. "Outcrops were sparse, but the flows were readily detected magnetically," wrote one chronicler. "Traverses of 6 mi a day were made across the cut-over timber, swamps, glacial lakes and muskeg of northwestern Wisconsin, by use of a sundial compass and pacing. A first-year man on the job earned no salary for the first month, and $45 per month for the following 2 months. After the first year one was qualified to take dip-meter readings, mapping the surface geology. The salary for this job was $60 per month and included a sleeping tent and meals of sorts." They worked seven days a week in the field and slept in the tent every night except Saturday, when they were allowed to make a night of it in the nearest town—within limits, of course.

Trouble with Allah

Despite the differences of opinion and hard feelings that resulted from the notorious 1925-26 International Geological Party, some good came out of it. In the early morning hours of October 14, 1927, the Baba Gurgur No. 1 broke loose with a bang. The well was the sixth to be drilled after the international expedition.

"At midnight of 14th/15th H.A. Winger, one of the American drillers, and his Iraqi crew went on shift," recalled F. E. Wellings. "They bailed out most of the mud left in the casing after cementing (keeping 500 feet of fluid at the bottom as a precaution against a blow-out), and the drilling tools were then run in. As soon as the bit cracked through the cement plug at the casing shoe, oil started to flow into the well. In a few moments it was blowing over the crown of the derrick to a height of about 140 feet—a typical old time gusher."

T.F. Williamson detailed events. "The driller ran to the boilers and extinguished the flames, and when daylight came the scene of the oil discovery was marked by a black pall of gas and oil vapours and the oil was flowing like a river down the Wadi Naft. At two o'clock in the afternoon the heavy drilling string was blown up into the derrick with a mighty roar.

"Three days passed before the control valve could be closed. Soon the ground all 'round the well was saturated with oil to a depth of several inches, and the slight breeze which rose daily was blowing some of it long distances. Men arriving from southern areas reported that it began to fog their windscreens as far as 10 miles from the well."

Five employees succumbed to gas poisoning. "Had there been one heavy rain storm all the hastily thrown up dams in the wadi would have burst and the thousands of tons of oil which they were holding up would have run downstream to the River Tigris, with consequences that would have been catastrophic." The well was finally closed in October 23—but not before the flow was estimated at 90,000 barrels per day. Citizens in nearby Kirkuk—which consisted of 800 mud-brick houses, a mere handful of shops, a few schools, no proper streets and no electricity—were nervous. The violent way in which the oil burst forth led many to believe that Allah was punishing them for their wrong-doing. Instead, it was just the beginning of extensive activity that would awaken the Middle East.

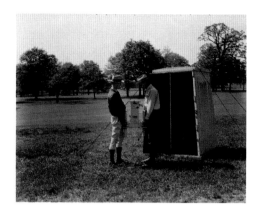

(Above)
The torsion balance was a major technological breakthrough in discovering oil around Gulf Coast salt domes. The torsion balance measured variations in the earth's gravity, which were used to indicate subsurface formations. Robert L. Kidd, board chairman of Cities Service of Bartlesville, demonstrates the machine in 1925 with the help of William F. Absher, Bartlesville, Cities Service geologist and founding member of AAPG. Cities Service Co.

(Opposite)
Cities Service used this Boeing 40B4 for research and development of fuels and lubes, about 1925. It was the same type of plane used in the first cross-country air mail. Cities Service Co.

The Acid Test

By the late 1920s, conditions were somewhat tamer in Mexico—at least everyone seemed to think so. Some wives were even willing to rough it with their geologist-husbands. When Charlie Hares took an assignment in Mexico, his young bride, Harriett Leonard, went with him. Harriett had been a teacher from the civilized city of Dallas, but she often accompanied Hares as he scouted the country.

On one occasion, the two got lost on a back road and stopped to ask directions. "All of a

Geologists pose at the Artesia field, New Mexico, in January 1925. Left to right: John Knox (Gulf), Grant Blanchard (Marland), Roy Hall (Gulf), Virgil B. Cole (Gulf) and Cliff Mohr (Marland). V.B. Cole.

sudden, we saw two men with long, thin mustaches approaching our auto," said Mrs. Hares. "I told Charlie to get out of there, and we went like blazes. Later we learned the spot was a bandit camp."

Bandits were not the only problem in Mexico. "The simplest food was often hard to get," wrote W.J. Archer, "and after a weary day beneath the scorching sun, slowly carving a way through the jungle, one's night was tortured by insect tribes from which no make-shift shelter afforded protection. To make a pathway through such territory, untouched for centuries, and without knowing what lay ahead, is an acid test of endurance and confidence."

The Hareses even tried outdoor living, in a tent near Monterrey. "The anticlines were beautiful," Hares noted, "but the living conditions were miserable." Flies, mosquitoes and dirt were everywhere. Both of them caught malaria. They took quinine for seven weeks before they could get to a doctor. Hares finally wrote Ohio Oil and asked to be sent instead to California. The company agreed, and the Hares pulled out. Mexico just wasn't worth the hassle.

Hares opened a one-room office in Los Angeles, with M.D. Woolery as manager and Elmer Bolton as landman. The area was a hotbed of activity. William F. Barbat was in the area for Standard Oil Company of California making a plane-table survey of the floor of the southern San Joaquin Valley. Just before his work was done, a dry-gas field—Buttonwillow—was discovered. There were several field geologists working the area, and it was not unusual for them to hear the wells come in, even when they were miles away.

Ottmar F. Kotick was at Devils Den, 50 miles away, when Milham Exploration Company's great gas well at Buttonwillow blew out and caught fire. He heard the sound and could see the fire. He decided to get closer. "Nearing the flame, the overpowering noise and heat represented a far greater force within the earth than I had ever thought possible," he wrote. "Cars parked 100 yards away from the well were burning. It was a most awesome experience. Later, after choking itself off, the resulting surface hole was huge and many feet deep."

By the time Hares arrived, many oil men thought that the good land had already been leased. Hares was not to be deterred. He immediately hired an airplane (similar to the one in which Lindberg flew to Paris) and flew over the San Joaquin Valley, where he could see the Buttonwillow gas field, its formations cropping out white on the surface. On the east side of the valley he spotted several small strike faults—probable traps. There he chose a well site.

He was right on target. Hoyt Gale, one of his old friends working the same area for Gulf, was exasperated at Hares' quick success. "You find the faults in an hour," he remarked, "and it took me all summer to map them."

In the jungles of Borneo, Orr and R.E. Dickerson pose with native workers at the Kari Orang Well, February 9, 1925. H.R. Cramer. Two days later, Dickerson shipped out. H.R. Kramer.

Whisenant (left), Dickerson and Orr donned their white tropical suits and pith helmets for the journey. H.R. Kramer.

Run Out of the Bean Field

Frank Albert Morgan also worked in the California fields in the 1920s. His wife, Helene, "always seemed to be available for field work when someone had to hold one end of the tape, or drive a stake for a wildcat well," explained Henry H. Neel. "In 1928, Helene did drive the stake for the first well in the Elwood field, which was the first 100-million-barrel field on the Santa Barbara coast...I have it on good authority that she was run out of the bean field by the truculent farmer while Frank was pacing off the coordinates of the location." Morgan was credited with "geologizing the area" and convincing Rio Grande Oil Company officials of its possibilities.

Locations and conditions on the wildcat wells were always top-secret. Morgan decided to try an unusual tack to get information back to his constituents. Frank, Jr., was a homing pigeon expert, and Neel would take crates of pigeons out with him to the wells and send messages back. "Soon one of the big companies equipped its field geologists with shot guns," Neel said. "Pigeons were fair game."

Boris Laiming mapped on the west side of California's San Joaquin Valley about 1926. C.A. Evans/Aminoil.

Riding Through an Oven

Fred Sutton was still in South America as field geologist for Standard Oil of Bolivia, investigating both Argentina and Bolivia. At times [and at a distance], he was taken with its beauty, and when he wrote his wife, Anne, he noted the clean, cool appearance of Bolivian towns—"adobe houses painted white & with red tile roofs. Viewed from the surrounding hills, they look clean and cool & refreshing, but when one is in them, one finds them hot, dirty, and depressing."

Sutton was in Bolivia as field geologist for Standard Oil of Bolivia and his job was to investigate Argentina and Bolivia. It was a merciless job. "The sun beat down out of a cloudless sky & it was like riding through an oven," he wrote after a trek to Lagunillas. "I arrived looking like a ripe tomato; even got sunburned back of my ears, and ached all over. Those muscles that have been dormant the past 5 months certainly felt the strain...I must have drunk 25 glasses of water, 6 cups of coffee, & 4 cups of tea, and it all stayed with me."

The expedition set out for camp over two ranges of mountains. "I have never seen so many bugs in one place in all my life," he wrote. "Garrapatas, mosquitoes, Meragui, and countless others. The one thing to be thankful for is that they all go to bed after dark, so that we can sleep in peace." Sutton was intrigued with river transportation in the back country. The Indians built "caballos," squared logs about six feet long which curved up at one end. "It is made of a certain kind of root that is just about as light as cork. They lie flat on this log, grip it with the knees & paddle with their hands. Unless one is accustomed to it, it is very difficult to do."

Sutton's men proceeded on a week's walking tour of the Rio Grande which almost did

Geologists pose at Wolf Camp in 1926. They are R.F. Baker, S.G. Gray, Philip B. King, Robert E. King, and A. Bruyere. R.E. King.

Dr. Paul P. Goudkoff examines core samples at a well in the San Joaquin Valley, about 1926. C.A. Evans/Aminoil.

them in. "The heat was almost unbearable, & the bugs frightful," he wrote. "We had a small boat which we used when we had to cross the river but the rest of the time we walked. My boots that I bought from Dr. Vasallo, have rubber soles, & I just about ruined my feet stepping from one hot rock to another. They are awfully sore on the bottom so I am going to stay in camp a day to rest. Yesterday was the hottest day of all. I got to camp first, drank about 4 qts. of water, then ducked my head & had a mozo pour a bucket of cold water on it…Singleton dragged himself into camp about 2 hrs. later asking for water, 'For C---- S---- traigame agua!' He then flopped on the ground and didn't move for about half an hour. Barale pulled in last of all, and called upon all the Italian saints to witness that no man could continue 10 days straight running without passing out." The group cheered up after a good bath and food. "Gus, the cook, is…a mighty good cook, & no matter what the occasion, he's right on the job, & has a meal ready in short order. He used to run a 'Quick and Dirty' lunch room in the States, so he can speed things up when he has to. He prides himself on not being an ordinary Greek, but a Spartan."

When Virgil Cole worked in Colombia, South America, in 1926, everything moved by mule pack trains. "There is no edible plant life in the jungle," he wrote. "We lived out of cans and boiled all water." Malaria and dysentery were rampant. Cole apologized for the poor quality of his photos. "It was nearly four months before I could get out and have the pictures developed and printed," he wrote. "With the heat and humidity, and only an Eastman Kodak camera, and used by an amateur, you can understand why some of the best shots never came out." V.B. Cole.

In 1926, oil men got together to discuss conditions in Colombia. Left to right: General Diaz, the revolutionary who claimed ownership of the area; Stev. Ghio, production superintendent of Gulf Oil; John Leonard, Pittsburgh, Pennsylvania, world oil developer who acquired and turned Barranca Bermeja over to Standard Oil of New Jersey; and I.A. Manning, Gulf office manager in Cartageña. V.B. Cole.

R.O. Rhoades graduated from Stanford University in the spring of 1922. He joined Gulf in 1926 as a geologist in South America. Rhoades later became vice-president in charge of all production, and chairman of the board of directors. He retired in October 1960, but died July 1961, from polyps on his intestines, which had been acquired in the rain forests of Colombia.   V.B. Cole.

(Left)
The Los Monos #1 discovery well in the jungles of Colombia was 400 kilometers inland from Cartageña. It was located in the center of about 100 acres of oil seeps and solid asphalt. Despite these conditions, the jungle growth was still thick. "The well burned for two weeks before we could finally keep the fire out," wrote Virgil Cole. "It burned out the bridge leading to the well. Fortunately the well only flowed by 'heads' 250 bbl per day."   V.B. Cole.

Geologists pose at Horquetta Camp, Magdalena Valley, Colombia, in 1926. Left to right: the native mule skinner; Martin, the instrument man; V.B. Cole, geologist; Oscar Hatcher, geologist; the native cook, and the native boys who helped them wade across the streams.   V.B. Cole.

Oscar Hatcher and Arch Gilbert take a break at Horquetta Camp in the Colombian jungle in 1926. Daniel (right) was interpreter and "body servant"; the other man was the Chinese cook.
V.B. Cole.

Cuebra camp, 1926, "one of our homes" in Magdalena country, Colombia.
V.B. Cole.

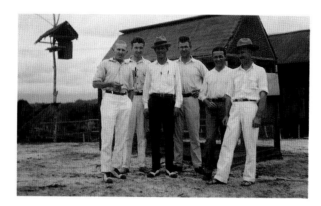

(Below)
Camp Aguas Claras was 400 kilometers "back in the jungles" of Colombia. The shelters were built on stilts and the area fenced to cut down on unwanted jungle visitors—scorpions, snakes and cats.
V.B. Cole.

(Left)
Six Gulf geologists gather at Aguas Claras, Colombia, in 1926. Left to right: (?) Martin, Grat Lynch, Oscar Hatcher, Ralph O. Rhodes, Arch C. Gilbert, and Virgil B. Cole. The men might be in the jungle, but they sported light shirts, white pants and two-toned shoes. V.B. Cole.

High Water and Barbed Arrows

About the same time, in 1927-28, William Argabrite, Walter K. Link and William Henry Wright began mapping the area east of the Colombian Andes from the Venezuela border south to Ecuador.

"They left Maracaibo by boat with 15 mules, and then by railroad, finally arriving at San Cristobal, Venezuela," Clarence S. Ross explained. High water in the Arauca River prevented them from moving into the Llanos (Colombian plains). As a result they worked their way up the Casanare River, and crossed the eastern range just below timberline at 14,000 feet. After turning south, they recrossed the range and reached the Llanos. During their work, they crossed the range seven or eight times, going all the way to the Magdalena River several times.

In July 1928, they returned from the Llanos and began work at Arauca. This trip depended entirely on riding and pack animals, and most of the food was gleaned from the immediate country and by hunting and fishing. "The settlers and Indians were cooperative largely through Argie's deft handling," Ross wrote. "There were no maps of the country and they picked up guides here and there who passed them along to other guides."

On another trek, Argabrite, Dr. Bela Hubbard of Standard Oil of New Jersey, and an engineer by the name of Briggs, investigated the eastern slope of the Perija Mountains, western Venezuela. The Motilone Indians were on the warpath at the time, which made the trip even more dangerous than usual. The group went by launch from the headquarters at Maracaibo, across Lake Maracaibo, then up the Santa Ana River to camp. As they traveled the river, the launch was inundated with arrows from the banks; they stuck in the boat's sides like bristles. When the men started out through the jungle on foot, warning arrows landed in the trail ahead of them.

Hubbard pushed 30 miles into the heart of Motilone territory, cutting trails and mapping formations. When he completed his maps and report, he sent an armed peon back to camp. Only a few hundred yards from camp, the peon was hit in the abdomen by an arrow. The geologists threw up log breastworks around the camp as the Indians surrounded them. For three hours, they were besieged. Then, from the direction of the river, they heard gunshots. A rescue party of 16 peons led by an American soldier of fortune made it through later in the day, and the camp was saved.

Other groups in the area were not so lucky. At one camp, the driller was shot in the back, the arrow piercing his lungs. The barbs were so ragged that the only way to pull them out was to shove strips of galvanized metal into the wound. Despite the attempt, the driller died.

In another jungle attack on a Shell Oil camp, one engineer was killed and two others were lost in the jungle. When a Standard Oil rescue party found their bodies, the hands had been cut off and the hearts cut out as trophies.

Bela Hubbard and geologist Gilbert P. Moore traversed eastern Peru along the Santiago

Charles Ristorchelli, son of the French consul, Maracaibo, Venezuela, worked for Sinclair Oil Co. in the state of Zamora. Ristorchelli spoke five languages fluently and had been U.S. educated. His father was French-Corsican; his mother German. "A tree fell on him and killed him after we returned from the field, April 1927," V.B. Cole wrote.

(Above)
"Jefe [chief] row" at the Venezuelan Gulf Oil Company camp at Maracaibo, Venezuela, in 1927.

(Above, right)
Men of the Apure Venezuela Petroleum Corporation (Sinclair Oil) at Pedrega, state of Zamora, Venezuela, January 1927. Bill Heroy, chief geologist, is second from left. The photo also includes Dr. N.H. Darton, U.S.G.S.; Arthur Darton, his son; J.E. King, an Englishman; Tom Hearne; Roscoe Reeves, U.S.G.S. and John Lycett.   V.B. Cole.

(Below, right)
Max L. Krueger poses on the Rio Guasare, district of Mora Paez, state of Zulia, Venezuela, in March 1927.   V.B. Cole.

River and the headwaters of the Amazon, thence across Brazil to the coast; they reached Buenos Aires in January 1922. Hubbard and Moore encountered Jivaro head-hunters—the Huambiza and Aquaruna Indians—whose weapons were blowguns with darts tipped with the vegetable poison curare. The two men managed to pacify the Indians by astute trading. The knives, cloth, thread and needles they gave the Indians may well have saved their lives. The Indians had mastered the technique for shrinking human heads, and Gilbert Moore made his way back to the states with one such head, "about the size of a large eggplant," according to Robert H. Dott. "He probably got the head by some more astute trading. It created considerable interest on shipboard, and also among the customs officers in New York, but was not confiscated as contraband," Dott said.

Lagunillas Field, Venezuela, in 1927. The "ark" which carried officials from the wells to other vessels rests in the distance. V.B. Cole.

Comic-Opera War

Fred A. Sutton was still in Bolivia late in 1928, working from the Bermejo camp. "Things seem to be picking up here & I have enough work to keep me quite busy," he wrote his wife, Anne. "It looks as though they intend me to re-check all the geology and all the well locations of this entire area. That will be pleasant work & I shall be able to return to my own camp every night, & not be subjected to making one in the bush...I'm praying that they get a darn good field here & will need a permanent geologist. In that case I think I could get the job, and they would surely build me a house here and let me bring my family."

The jungle was always filled with surprises, and Sutton was confronted with an American who drifted into camp unexpectedly one day. "He must be about 40 or 50 yrs. old," he wrote. "I hired him more out of pity than anything else. He walked from Tarija, a distance of about 200 miles. He lost the trail and got into the Bermejo River, & followed it downstream until he reached this place. He was 7 days without food and was certainly a sight when he arrived here. He had worked in Colombia for 3 yrs., and knew a lot of people I know."

By December, Sutton had bigger troubles than itinerant drifters. Bolivia and Paraguay were at odds. "I do not know the exact status at the present moment, but rumor has it that war has actually been declared," Sutton wrote. "It will not affect us here, because to get here a part of the Argentine must be crossed, and Argentina would not let any foreign troops cross her territory in time of war. I understand that farther north in Bolivia, on the main road...the Bolivians have appropriated 700 company mules and 3 motor trucks. That is just about half of the mules we own. I cannot swear to the truth of this. If there is a war, I am inclined to believe it will be of the comic opera variety...[Bolivia's soldiers] are for the most part illiterate, half-breed...They are forced into the service. They have no idea what it is all about, & the word patriotism does not exist for them...The soldiers they have in

To cross the Orinoco River in Venezuela in 1929, the geologist's car was loaded on a sailboat. H.M. Kirk.

Rolf Engleman found that a hammock rig kept insects out pretty well—as long as it was closed up well. One night, however, he forgot to close it sufficiently, and he was bitten by a scorpion. "I slept with my boots for a long time," he said. Rolf Engleman/Mrs. Rolf Engleman.

W.D. Chawner worked with a bandanna over his head to protect against biting flies and mosquitoes, as he made a plane-table survey along the railroad from Puerto Wileber to Puerto Santos, Santander Province, Colombia, November 1927. W.D. Chawner.

Four weary geologists pose in Cuevas Pintas, Coahuila, Mexico, February 26, 1927. They are left to right: H.M. "Howdie" Kirk, W.A. "Bill" Clark, Linn Farrish and Tom Stipp. H.M. Kirk.

Villamontes do not even have shoes to wear...They may bring soldiers of better class from the highlands, but if they do, they will die off like rats when they come down to the Chaco, where the fighting will necessarily take place."

The trouble was over a vast alluvial almost featureless plain which ran east of the oil fields to the Rio Parana. "Transportation will be exceedingly difficult, & fighting will be entirely of the guerrilla type," he wrote. "If there is a war, I do not think it will last long, & I think the Bolivians will get the worst of it. It is quite an open secret...that the company's oil activities and the possible existence of oil deposits in the disputed territory have much to do with these countries wanting those lands."

By January 1929, Sutton was more concerned with the weather conditions, which were wreaking havoc on his camps. "We are indeed in the rainy season here. It began to rain last night about 10 o'clock & did not stop until noon today...The river rose higher than it has risen since this camp was started 7 yrs. ago...Large trees and great heaps of rubbish floated serenely by. It was still rising when I got up this morning, & the peons and their families who live near the river bank were all busy moving back to higher ground. It flooded them out completely. Men, women, and children were all wading in & out of their houses carrying beds, chairs, chickens...The women, as usual, did most of the work. They made countless trips back and forth in the heavy rain, & with the water lapping about their waists."

Sutton's engineer, Franco, was working upriver about 10 miles north at the time. "He was camped on the Rio San Telmo...His camp was located close to the river & he was flooded out. He saved his maps and notes, but lost all his food & other things. He & his men were 6 days without anything to eat except berries, leaves & wild honey. One of his peons & four of his mules were drowned in the river. One peon was sick & had to be left behind with someone to look after him. When Franco reached the Rio Bermejo, his peons refused to go back for the sick man, & he had to drive them back at the point of a gun."

Conditions did not improve much. One of Sutton's peons was drowned while crossing the Rio Bermejo, and it was four days before they found his body. Ten days later, the mail man reported that the mule that carried the mail had fallen down in the river, and the mail had floated downstream. Fortunately, the mule had gotten out all right. The precious mail had not fared so well.

Flying High

Early attempts at aerial photography in South America were only partially successful. The Bolivian acreage was mapped, but the plane missed the crestal areas of some anticlines. "A composite mosaic put together from these photos...was horrible," wrote George Harrington. "Small areas were left out or repeated in the mosaic."

Robert B. Moran, a Los Angeles geologist, was working in Peru also using aerial photography. Moran had gone to Peru to study the possibility of building a railroad across the Andes to connect the Pacific Coast with the Upper Amazon. He and geologist Douglas Fyfe went from Lima across the Andes by truck and automobile to San Ramon, where they transferred to an Army trimotor airplane.

Moran was one of the first to use airplanes succesfully for geologic reconnaissance and mapping. On one flight, the pilot flew over an elliptically shaped corona of trees and shaded vegetation which drew Moran's attention. He requested that the pilot fly over it in several directions as he studied it. When he returned to camp, he organized a party with Fyfe to investigate the area. They applied for and obtained a concession to the site, which was about 20 miles long and eight miles wide. (It was several years before the area was developed, but the Ganzo Azul oil field was brought in on this dome that Moran had identified.

A reporter once asked Moran how he recognized an anticline. His laconic reply was, "How do you recognize a cow?"

Getting equipment from one side of a stream to another could be a problem. When Lewis Weeks was surveying the rivers of the Amazon headwaters in northeastern Bolivia, access to the region was via a 400-mile, three-week mule ride from a camp in northern Argentina. His train included 28 mules. To keep his theodolite dry, he had it transferred across the river on top of a native's head. L.G. Weeks/AAPG.

(Opposite)
Phillips Petroleum established one of the first hydrocarbon research laboratories. Some scientific geological work was carried on in the lab which was located in the new office building in Bartlesville, Oklahoma, in 1927. Cities Service Co.

(Right)
In 1927, the Rachel No. 7 blew out at White Point field, San Patricio County, Texas. "The well was being drilled by Saxet Gas Co. and caught fire from static electricity caused by tiny rust particles in the gas flow massaging the casing pipe walls from water injected into the dry gas sand from a relief well drilled nearby," W. Armstrong Price wrote. "The crew fought the blowout for 110 days until the cause of the small blue glow at the mouth of the pipe was discovered and no more water was injected into the gas sand. By that time, the crater was 75 feet deep. Clay on the crater walls was glowing glass until it was cooled down by water. W.A. Price.

(Above)
In March 1927, the executive committee of AAPG posed at the 12th annual meeting. From left to right: (standing) David Donoghue, secretary-treasurer; John L. Rich, editor; Luther H. White, vice-president; (seated) G.C. Gester, president 1927–28; Alex W. McCoy, president, 1926–27.

(Above, right)
AAPG headquarters staff was growing in 1928. Anna D. Whalen and Margaret Shelton manned the Underwood typewriters. AAPG.

(Below, right)
In 1928, J.P.D. Hull, AAPG business manager, worked from a roll-top desk in an office in the Wright Building, Tulsa. AAPG.

Students at a summer field camp in Utah load up the truck bed. George Hansen, chairman of the department of geology, Brigham Young University, stands in back. Brigham Young University.

(Right)
John Fitts examines fossil Callixylon logs in the Ada, Oklahoma area, October 1930. Fitts started as a water boy for the U.S.G.S.'s original survey of the area. H. Miser.

(Above)
Burton E. Ashley posed near Minong, Wisconsin, while working for the Wisconsin State Mineral Lands Survey in the summer of 1929. B.E. Ashley.

(Below, right)
Professor W.A. Tarr, University of Missouri (Columbia) investigates Precambrian rocks in southeast Missouri about 1928. Tom Freeman/University of Missouri.

Hard Times in the Oil Patch

When the New York Stock Exchange opened on the morning of October 24, 1929, stock prices began a dramatic drop. Before noon, panic set in, and the stock market ticker tape fell further and further behind. The Exchange appeared relatively stable the next day, but four days later, prices started another downturn.

Panic increased on October 29. More and more shares were dumped on the market. Thousands of investors watched their fortunes disappear.

The stock market crash of 1929 was a symptom of conditions that already existed across the country. To men and companies in the oil business, it was just one more difficulty. For months, geologists had continued to identify new fields, and companies had continued to bring them in. But aside from the beginning glow of success, the overall enthusiasm was lacking.

The problem was that petroleum production had increased beyond demand. The Midwest area was particularly hard hit. When the Oklahoma City oil field was opened in December 1928, excess crude flooded the market. It was compounded by a production revival in the Greater Seminole field, which was just hitting its stride.

As times grew tougher and leaner everywhere, men who had only dabbled in drilling and leasing increased their efforts in hopes of striking a fortune and alleviating their current money problems.

Others who had always been drawn by the dazzle of oil were also still out searching for the reservoir at the end of the drill. One of those was Columbus Marion "Dad" Joiner. Joiner was no geologist, but he had a knack for picking the right spot to set down the drill. He had found the first oil in Oklahoma's Cement pool (although geologists Frank Butram and Jerry Newby followed closely with a gas well). He had also worked in the early Ardmore and Healdton fields. Even then, he had been eyeing east Texas.

H. Dow Hamm, district geologist at Ardmore for the Roxana Company, had been approached in the mid-'20s. "My secretary was a very nice little girl named Ella Springer," Hamm wrote. Ella lived at Joiner City, named for "Dad" Joiner, located on the old Ringling Railroad west from Ardmore to a few small whistle stops. The Joiners had moved to Ardmore after the Healdton boom had settled down a bit, and Mrs. Joiner had opened a boarding house. Miss Ella's mother was well acquainted with Mrs. Joiner and felt her daughter was safe living in such an establishment as long as she was under Mrs. Joiner's care.

"Miss Springer brought a 'geological' report to my office and said that Mr. Joiner would sure like to have me read it and let him know what I thought about it," Hamm wrote. "The report, with thin paper cover, was composed of only two pages. The first page was the text, and the second page was an oil [field] map of the United States. The map showed the outstanding fields of the United States at that time....The author of the report, one Dr. Lloyd, who, I believe, claimed to be or was a veterinarian, had projected lines from all of

(Above)
R.P. McLaren and Bill Harvey worked in the Delaware Mountains in west Texas, 1928. E.W. Owen/Mirva Owen.

(Left)
Donald McArthur digs up a nest of newly laid turtle eggs on the Ermitano River, Colombia, in 1928. "If fresh, they make a fine omelet," W.D. Chawner wrote. W.D. Chawner.

these major oil fields to a point of intersection in East Texas. I do not remember the preamble to the report, but I do recall the last paragraph of the text, which went something as follows: 'These trends from all the major oil fields in the United States, intersecting as they do at this point in East Texas, create a situation known as the apex of the apex. A situation not frequently found elsewhere on this earth.'

"After reviewing this report and the obvious lack of technical background, I had the unusually good judgment to tell Miss Springer [truthfully] that I really didn't know much about that part of East Texas, and...therefore would neither encourage nor disillusion Mr. Joiner. The young lady thanked me, and she took the report back to Dad Joiner."

Joiner was more optimistic, however, and in 1930 set up to drill in east Texas. He leased 10,000 acres between the Sabine uplift of northwest Louisiana and the faults which curve around east and northeast edges of the central uplift of Texas 60 miles to the west. There was already production on the uplift to the east and along the faults to the west, but geologists had thought there would be a dry area where Joiner proposed to drill.

Joiner drilled anyway, however, and brought in one of the biggest oil fields ever. It started another flurry of activity. Countless operators drilled more than 4,000 wells within a short 16 months after the new field came in. Oil gushed up at one million barrels a day—eight times as much as the entire world then consumed each day. Town-lot drilling was frenzied as men searched for oil in their own back yard. Crude oil prices dropped from $1.10 a barrel to 10 cents a barrel in only six months. One oil operator grumbled loudly when he paid 15 cents for a bowl of chili in an east Texas cafe because it was more than he made out of an entire barrel of oil.

When Sidney Powers visited east Texas in mid-1931, the area had 612 producing wells with a daily average production of 351,000 barrels. At that time 600 to 700 new wells were being drilled each week and 150 new locations were being staked. Powers wrote Everett DeGolyer, "I spent a day in east Texas and got one idea which I think is sufficiently important to transmit to you: The companies are playing East Texas for profit after 1932, and they expect a very long life and profitable yield after the flush is off. They do not expect a profit in the immediate future. I think we should consider this point of view for the Amerada, if we are going to plan to make money in the production end of the business or to amass holdings which we can sell about 1933 at a considerable profit. I believe that the chances are far better that no large oil field will be found in the next five years, than that another field will be found; and I think that we should keep an eye toward the return of prosperity in the oil business, and not be over-influenced by the present gloom which will last into 1932....

"I am told that a number of good trades can be worked out in east Texas now and that we might be able to find some properties which, like our own, would not need intensive immediate development. If so, this is a possibility for a very attractive play.

"I am told that royalty prices have risen to about $400.00 an acre, and that very little

By 1928, the Magdalena River could be traveled by SCADTA seaplane instead of the slower river boat. It was a toss-up which was more comfortable or safer. W.D. Chawner.

J.E. "Brick" Elliott (right) and friends pose at the Duck Club near Watts, California, in the fall of 1928. "I had the fuel tank space forward of Pilot converted to 4-place passenger cabin," he wrote, "snug but adequate."

trading is going on because the owners think that they can ultimately make more than this and they are holding [out] for future sales."

The production rates in Oklahoma and Texas were alarming, and efforts at self-regulation went unheeded. Oklahoma Governor "Alfalfa Bill" Murray tried to persuade oil men to adopt proration, in order to lessen the supply and thus increase the prices. When they would not, martial law was declared in both Oklahoma and Texas, "until," as Murray said, "we get dollar oil." Murray ordered the National Guard into Oklahoma's oil fields to enforce a shutdown of all wells producing more than 25 barrels daily. Texas governor Ross Sterling eventually followed suit, although not until crude oil prices had plummeted to five to seven cents a barrel. Fortunately, the measure worked. Within days, crude prices rose to nearly the dollar mark, and the ban was lifted. Oklahoma adopted a proration schedule that became a permanent feature of its oil-production program.

Goose Chase Blues

Martial law in the Midwest did not at first keep companies from putting geologists in the field; it only prevented them from drilling in certain areas. They simply shifted areas of emphasis.

One young man, disgusted with the summer and the slim pickings in Arkansas, wrote

Dollie Radler at Amerada on July 3, 1931, "Am still convinced that this isn't the proper season to do geological work here. The thermometer has climbed above 100 every day, the highest temperature being recorded was 104 [degrees] Wesneday [sic]. The record was taken in the shade, so no telling how hot it is in the sun. Was 110 in front of the hotel at Clarksville last Sunday...Will continue to look for structure, but have the feeling that this business will turn out to be a wild goose chase."

Some geologists were willing to do anything to find oil, and Wallace Thompson must have been one of them. He became a true subsurface geologist. John Freeman called him the first "talking electric log." A wildcat well in Cottle County, Texas, was being drilled with cable tools in a large hole, but the bit broke off at a depth of only 65 feet. The driller decided to go down the hole. He saw the slanting wall of a crevice in gypsum which was deflecting the tools, so he cut a bench with a hammer and chisel to keep the hole straight. The well was about two miles from the outcrop of key beds which had been used to map the structure, and Thompson wanted to see the formation. They proceeded to drill a few feet deeper; then Thompson put his foot in a loop in the line and was lowered to the bottom. He inspected the walls of the hole with a flashlight. When this view confirmed his ideas of his structure, he phoned Chief Geologist [F.H.] Lahee at headquarters. Lahee was equally excited and reported to the big boss, J. Edgar Pew. Pew was less enthusiastic. "We have enough trouble fishing for tools," he remarked, "keep the geologists out of those holes!"

As the Depression grew more widespread, oil companies began to cut back, and the ranks of the unemployed geologists grew. In an effort to help young graduates on the West Coast, Olaf Jenkins, head of the California Division of Geology and Mines, created a series of field mapping projects.

"The pay was miniscule and the areas selected were pretty brutal, even by California field standards," explained John Charles Hazzard. Jenkins assigned the young men such projects as the Mojave Desert, east and south of Death Valley—the Marble Mountains, Providence Range and the Nopah Range. They were not only to complete geological examinations and tests, but they also had to make their own maps.

Hazzard was one of the 'lucky' young men. The party of two or three young geologists lived, ate and dressed with the frugality of desert Indians. The Providence Range was mapped mostly in the summers, and the temperature rose to 120 degrees. Water was at a premium. When they could no longer carry enough for a day's work, they gave up summer work. They mapped and surveyed in winter when they could drink meltwater from the snow.

Because roads were abominable tracks, or nonexistent, they moved from spike camps set up wherever they could coax their old jalopy up to the foot of a mountain. Cans of water and stacks of spare tires usually adorned the old cars in every available space.

"At that time," explained William H. Easton, "it was generally believed that a great hiatus separated geologic time into two increments and life probably originated during the

Rolf Engleman holds a hammer from which hangs a stringer of black viscous oil, October 1931. The seep was 1800 meters northwest by north of Guamutas, Colombia. W.D. Chawner.

interval of hundreds of millions of years of unrecorded time. But John [Hazzard] described thousands of feet of rocks that occupied that supposed gap in the geologic record." The work Hazzard performed on the Providence and Nopah ranges was a major contribution to geology, and changed attitudes toward the geologic process.

Scotch on the Rocks

The Depression of the 1930s was worldwide. There had been enough geologists active in Canada in the late 1920s, however, to form the Alberta Society of Petroleum Geologists. When the larger exploration companies pulled out, though, many geologists were left high and dry. Some were given temporary jobs by the sympathetic Provincial Government, although not all these positions were directly related to their profession. Many of the men who worked in foreign countries found it a hard row to hoe, but whatever the job might be, it was better than no job at all—which is what awaited them back in the states.

In 1932, Ed Owen received a letter from a friend working for N.V. Nederlandsche Koloniale Petroleum Maatschappij (NNKPM) in Tratak Boru, Sumatra. "As you see, old man procrastination has been with me but I got rid of him the first of the year," the geologist wrote. "Your welcome letter was received months ago while I was in Borneo, but all the interesting wild men & women, together with the old man mentioned above, prevented my answering sooner.

"You may doubt it but we have a depression here also and I contend that it is worse than in the states. A few days ago four Chinese...offered to work for me as coolies at 80 cents Dutch (32 cents gold) per day. When a Chinaman finds it impossible to make a living in business, there is no business...

"You made a few nasty cracks about drinking before and after breakfast. I will admit that I have had a few drinks since coming to the D.E.I., but I am not the man I used to be. My capacity, compared to some of my English & Scotch friends, is nothing to talk about. The depression has hit them hard and most of them have had to limit themselves to 1 qt. per day. Of course some of the old timers cannot get along on that, so they cut down on food rather than whisky. My personal record is this: from Apr. 1, 1930 to Aug. 20, 1/2 quart (total not per [day]), from Aug 20 to Nov 25,-more. Since starting this job I have had four or five drinks and from here I can see 1 qt Golden Wedding (Rye), 1 qt Old Log Cabin (bourbon), 1 qt King George (Scotch). If I had someone to drink with I could be tempted. How are the 'bootleggers?'

"We, the N.K.P.M., have developed a hand drilling and coring apparatus and have entirely done away with pit digging. We get a core 6″ to 8″ long and 2″ in diameter. It is properly oriented and providing there is stratification, the dip and strike can easily be read.

A crew of three men can drill six holes eight meters deep in one day and the coring crew, five or six men, can take 30-40 cores per day. The work goes faster than the old pit work and it is quite satisfactory. It is possible to drill holes 10 to 15 meters deep. My average on this job is about six. I have three surveyors working, and they cut and survey about two kilometers each, per day; and as the core holes are spaced 100 meters apart you can see that a large crew is necessary, and consequently the expense is high even with cheap labor...

"Possibly you have heard that the Abilene State Bank failed. Well, I went broke the same day; consequently, I am hoping to stay an extra year and then probably come back for another contract. I believe that all the American geologists want another contract, with the possible exception of optomistic H—but the Co. will make no promises or statements. The local chief wants us to return but New York is saying nothing. If we go home I will have to start digging ditches, if there are any ditches to be dug, and providing too many other better men are not after the same job, for I infer from your letter and other sources that the oil business can support no more men."

(Following page)
Geology students with Oklahoma A & M traveled in style in 1931—with MK & O Coach lines. Blakey Group.

Parties and skits were typical extracurricular affairs with Cities Service's Doherty Girls Club, Bartlesville. The photo was taken April 7, 1930, after some special but unidentified presentation. Cities Service.

An oil derrick and tanks perch atop a crest north of Alhambra Avenue, Los Angeles, in 1931. The crest is of a small fold in thin-bedded Monterey Formation.
M.N.Bramlette/U.S.G.S.

Salt pans on the major thrust line in the higher Zagros Mountains were supplied by early Cambrian salt.   N.L. Falcon.

The Gach-i-Qaraquli well #4 was drilled in the winter of 1929/30 using the same cable tool rig that was used when the major Gach Saran oil field #3 was drilled in 1929. "Geophysics was only just being developed at the time," wrote N.L. Falcon, "and both wells were located on geological surface guesswork. It was a number of years before the shape of the hidden Asmari limestone anticline below the confused Lower Fars (now Gad-Saran) formation was defined."   N.L. Falcon.

(Above)
Atlantic Refining Co. geologists used one of the private railroad rail cars on a large American-owned sugar plantation, in Cuba in 1932. Left to right: W.D. Chawner, Norman Weisbord and Rolf Engleman. W.D. Chawner.

(Left)
Atlantic Refining Company officials investigate the Valentino Oil Seep, San Juan de Wilson, Cádenas area, Cuba, January 23, 1931. Left to right: W.D. Chawner, geologist; Dr. R.E. Dickerson, chief geologist, Fureja?, J.W. Van Dyke, president, Atlantic, and W.M. O'Connor, vice president, Atlantic.   W.D. Chawner.

Depression Down Under

Australians hoped that petroleum development might be a solution to the economic problem and worldwide depression which had spread even "down under." In a report for the Australian parliament, government geologist W.G. Woolnough wrote, "The discovery in Australia of an entirely new source of wealth, providing an avenue of employment for large numbers of people, would do more to bring about a return of prosperity than almost any other thing."

The trouble was that many Australians felt no oil existed in their country. Worse still, the British who ruled the island continent did not think big enough. "British Governments and business men rarely have the breadth of view in financial matters to count the cost, in advance, in sufficiently big figures, and tend to 'nibble' at the job," he wrote. "They look upon 100,000 pounds as an enormous sum; but few great oil fields have been discovered for less than ten times this amount, and years of patient research are called for. British operators tend to get 'cold feet' as a result of initial financial setbacks." This meant that given limited resources, Australia was faced with limited information on which to act.

Woolnough recalled one instance in which he had visited a "certain oil permit, of which a 'map' was printed in the prospectus. During a Sunday morning walk of less than a couple of miles I came across conspicuous outcrops which had a very important bearing on the problem, which had evidently never been visited by the 'geologist' who had prepared the 'map.'"

Such attitudes put Australia far behind other, more progressively oriented countries. "Until Australians come to realize that the preliminary geological mapping of a large permit occupies months and not days, there can be no really wholesome appreciation of the seriousness of oil search." Woolnough recognized the good work of the Victorian Geological Survey and one or two oil companies whose works compared favorably with "the type turned out as a matter of everyday routine in the United States...Within the last year certain other companies have taken this lesson to heart, and there promises to be all-round improvement in the matter of geological mapping for oil purposes in the near future."

Woolnough was concerned also with those who purported to be geologists in order to swindle investors. "Generally as a result of the seductions of the bogus 'geologist' or of the 'doodle bug artist' (alias 'geophysicist' in some instances), small communities are fired wih the patriotic object of raising their hamlet, in one act, to the proud position of becoming a city. Groups of farmers and local tradesmen contribute their hard-earned savings, widows and orphans withdraw their funds from the safety of gilt-edged securities to reap the fabulously rich and 'certain' reward, 'guaranteed' on the 'honour' of the promoter. The result is nearly always the same, as it has been in too many cases in this country, disappointment and ruin for the many, wealth for the individual or the few. History is not without cases of entirely unsuspected successes attained, through sheer luck, in this type of

Philip B. King looked every bit the "typical field geologist" when he posed in west Texas about 1932. R.E. King.

operation. One of the great fields of Texas...is said to have been started as a downright swindle. The first well blundered into an oil pool...

"If investors would insist upon the production of a detailed geological map as a prerequisite to contribution to the funds of an oil company, there would be a document sufficient to prove or disprove the qualifications and honesty of the person who has produced it...It is time that the public demanded an equally satisfactory documentary proof of the competence and honesty of the 'geologists' employed by them."

Midst of a Revolution

In 1931, Standard Oil of New Jersey sent Carl B. Richardson and Ruthven Pike to Bogotá, Colombia. The two had been in Peru and had to make their way to Colombia via Lima and New York. All field equipment and accessories were shipped down the Amazon to Para [Belem] and transhipped from there to New York. In New York, Richardson, Pike, gear and instruments were loaded aboard a United Fruit steamer to Barranquilla. Three weeks later, via river boat and railroad, they arrived in Bogotá.

Pike was assigned the southern half of the concession, and Richardson drew the northern half. They moved by foot or mule, measuring distances by transit, stadia hand level and occasionally by careful pacing. It was a large job, and the deadline neared long before the work was completed. They worked on, into the rainy season, as swollen rivers grew more dangerous. Even the mules could not swim against the currents carrying loaded pack saddles, so they had to be unloaded in order to be able to cross with any degree of safety. Only the larger, heavier oxen made the river crossings with comparative ease.

While Richardson and Pike were slogging their way through Colombia, the same company sent Kessack White to Brazil. White was used to hard conditions, but this time he was more fortunate than the crew in Colombia. He not only had a horse, but an auto at his disposal, as well.

"We mapped a large part of south Brazil topographically and geologically," he wrote. "Our tools—automobile speedometer, horse and foot pacing, Brunton compass and aneroid barometer...There were a few railroad lines that served as a base and bench-mark-elevation control. Roads, where passable, were traversed by auto; two to three aneroids [barometers] were carried and read at each locality."

Meanwhile, Victor Oppenheim had returned to Buenos Aires from an expedition in southern Argentina. Oppenheim was a seasoned traveler, a Latvian, the son of a world-traveling civil engineer. He had been educated in China and was a veteran of Swedish expeditions into Mongolia and Tibet. He picked up instructions for his next expedition, this time to the Cordillera del Condor, a little-known mining region in the northwestern Argentine Andes.

"As I completed preparations for departure, news flashed through Buenos Aires that cavalry troops under General Uriburu were converging down the main streets of the city toward the Casa Rosada, the presidential palace," he wrote. Oppenheim stepped to his hotel window as a regiment of Chaco horsemen, the toughest of the Argentine cavalry, rode by, charging at full gallop down the Calle Florida.

"People ran in all directions," he wrote. "Some looked for refuge in the stores and shop entrances. Others clung to the walls of the buildings trying to avoid the onrushing wave of horsemen. Many bystanders were trampled by the horses; others were wounded or killed."

Oppenheim listened to the machine guns and the rifles as Uriburu's troops took over the Casa Rosada. The sporadic shooting from the Casa Rosada continued late into the night and the next day. Streets were deserted. Although few left shelter, many were shot by stray bullets or by trigger-happy sharpshooters. Once in a while, someone would dash along the empty streets, waving a blood-soaked shirt or handkerchief.

"I could not understand at the time who was for or against the revolution, nor was it clear who had won or lost, or what was the cause or object of this military coup d'etat," Oppenheim wrote.

Finally, he decided to venture out. About midnight, he walked slowly down the deserted Avenida de Mayo, toward the Plaza de Mayo. "About two blocks ahead of me a man appeared, racing across the street and shouting something in a very excited voice. A few short, dry bursts of machine-gun fire were heard as bullets whizzed along the avenue, fired from the Casa Rosada at the end of it. The running man fell and lay quiet. More shots were heard and it seemed that some of the bullets passed quite near me! I turned back quickly and, clinging to the walls of the buildings, to offer as little a target as possible to the invisible occupants of the Casa Rosada, I returned to my hotel, shocked by the sight of the fallen citizen."

More than a hundred people died during the revolution although the exact number was never revealed. Oppenheim was one of the lucky survivors.

Victor Oppenheim surveys the Abaete River Valley of Brazil in 1932.
V. Oppenheim.

Coca for Pacha-Mama

After the military coup, even the formidable Andes seemed safer territory, and Oppenheim set out on his planned expedition. He passed the headwaters of the Rio Piedras and the Rio Canas, and proceeded on to the Andes with their barren, rocky peaks.

"Our narrow, slippery path wound up the face of the sheer barren cliffs of the Cumbres Grandes," he wrote. "It was barely wide enough for one animal." Oppenheim was quick to recognize the value of the native mules, "the most sure-footed animals in the Andes. They evidently have a sense of balance all their own that tells them where to put their forefeet and where not to step under any circumstances, which gives them a unique sure-footedness.

A mule will walk on a dark night along the most dangerous mountain trails, sniffing the ground it steps on, barely seeing, yet, at the first place where it is not sure, the mule will stop, and unless pulled or pushed over that spot, it will not move one step."

As they climbed higher, the geological formations began to change. Oppenheim grew excited and wanted to stop to study the rocks along the ledge. His mount refused to stop, and stubbornly continued to climb in spite of all urging. "With a several thousand foot drop on one side of the foot-wide trail, and a sheer wall on the other side, no good mule should ever be stopping to contemplate the landscape or wonder about the rocks at the bottom of the precipice," he wrote.

Late one evening, Oppenheim reached a small flat platform on the eastern wall of the Cordillera Oriental, about 14,000 foot above sea level. It was well below freezing. "We had scarcely unsaddled our animals when darkness fell," he wrote. Since there was no space on the ledge for both mules and tent, the men spent the night huddled against the mountain wall.

Oppenheim was eager to climb the Cerro Alto Morado, a forbidding peak of about 17,000 feet which towered above the surrounding mountains. "We left at dawn one morning, my assistant, two Indian porters and myself...Our mules were slow and appeared depressed, because of the high altitude. They don't fare very well at excessively high elevations and so we left them at an adequate spot.

If it couldn't be carried by four-legged transportation, Victor Oppenheim usually did without it. Oppenheim and his white mule explored the chapada of Mato Grosso in western Brazil, in July 1932. V. Oppenheim.

"All this time, our Indian porters were in the lead. They were climbing at a steady and even pace, which I thought was a little too fast for this altitude. Their walk was as light and as agile as that of the guanacos and vicuñas…Suddenly, my assistant, who was climbing far behind us, called to stop…He appeared to be in distress, pale in the face, and very ill, with blood streaming from his nose. This was *soroche*, or the high-altitude sickness of the Andes, which in some cases, should the patient lose consciousness and the heart be affected, can be serious."

Oppenheim took the assistant further down the mountain, left a guide with him, and proceeded. The other guide, Huanco, was extraordinarily fast and light. "Eventually, I could no longer continue at his pace. I had to stop frequently and breathe deeply at every pause. During one of these pauses, my Indian friend…took some coca leaves from a little bag attached to his belt. He carefully placed them on a large boulder and put a smaller stone on top, and then murmured some Indian words." Oppenheim asked him what he was doing. "This is for Pacha-Mama [the goddess of the mountains]," the Indian replied. He explained that the air was rare and hard to breathe—events which Pacha-Mama sent to make the travelers go back. The valuable coca leaves were to appease the goddess.

"I was getting shorter of breath and my pauses were more frequent," he wrote. Finally, they reached a high wall some thousand feet below the top of the peak. The Indian scurried up to the first ledge and clung to the wall with the toes of one foot wedged in the rock hollows, while the other hung freely over the precipice. "I had not expected to face such obstacles," Oppenheim wrote, "and had left the rope below with my assistant. But I doubt that it would have been of much use to me, as my Indian friend had no notion of using a rope in the mountains. All I had to help me up that wall was the pick-end of my geological hammer."

Oppenheim jammed the hammer into the crevices and hung on by it to any ledge that seemed capable of supporting his weight. "I did not dare look down. I knew there was nothing there. My eyes and all of my energy were concentrated on that wall in front of my face and above my head. Each panting breath was a struggle; there was never enough air…With my heavy boots on, I could hardly follow this fellow, with his light gait of a mountain animal, his blind faith in the protection of Pacha-Mama, and his coca leaves. But now the bridges were burned behind me. I could not possibly climb down from where I was clinging. I had only one way and that was to climb and pull myself up and ahead at all costs, at any cost…Suddenly, when my legs felt particularly unsteady, and my boots seemed to be loaded with lead, I pulled myself up and crawled out on a wide, rocky ledge…When I cursingly asked him, after a long pause for breathing, why he did not warn me of this dangerous climb, he looked at me surprised, and said, 'But this is the shortest way, Señor; the other way is long.' "

Victor Oppenheim was in charge of one of the first exploratory wells drilled for Compania Nacional de Petroleo in the state of Alagoas, Brazil in 1932. V. Oppenheim.

Mountain High

On several occasions, Oppenheim tried the coca when climbing in the high mountains. He found the leaves slightly bitter, pleasant, and rather refreshing. "After several hours of chewing, I noticed that I could climb longer without tiring and was not as short of breath as one would normally be at altitudes above ten thousand feet...I could easily understand why the Indians prize their coca so highly."

After several weeks in the Quebrada de Humahuaca, valleys of red sands and dust, Oppenheim's group ascended the Serranía de Zenta. They crossed at the Abra Blanca at about 14,000 feet in elevation. Snow was falling in the mountains, and the trail was barely visible. They camped near a rushing mountain torrent, in the lee of a high peak. The mules had eaten all the green barley and alfalfa the guides had packed and had to pass the night without any food, in spite of the strenuous climb up the mountains. The high camp was extremely cold, and they were glad when the light of day came. During the night, the wind had carried off some old discarded newspapers used for wrapping. The hungry mules were nibbling at the ragged papers. Oppenheim fished out pieces of raw brown sugar from the men's food supply and gave a piece to each mule. "This they ate avidly, and their dispositions and energies visibly picked up," he wrote. By the time they reached Colanzula later that day, they were able to secure enough barley to last the animals several days. But the mules had worn flat their horseshoes on the sharp trails of scree and rocks. The group had to stay long enough to have their mules re-shod.

Confronting the Headhunters

Oppenheim joined the Argentine government petroleum company (YPF), to study the north Argentine oil fields. He lived in the main field camp, which held more than 200 employees.

Rules were strict. No women—wives or otherwise —were allowed. Drinking was also forbidden, but those whose thirst was unquenchable usually managed to find excuses to head to the neighboring village at least once or twice a day.

When Oppenheim moved out into the Gran Chaco, it was the rainy season. Oppenheim's team was plagued with the problems that all field geologists encountered. There was trouble between the Gauchos and the Chinese cook. Coral snakes and fer-de-lance were abundant, and Oppenheim once shared a bed with a coral snake. The campsite was flooded out by rising waters from the rain. Oppenheim took it matter-of-factly.

When Oppenheim headed out for western Brazil, little was known about its geology. The state of Mato Grosso bordered Bolivia, Paraguay and the Amazon Valley; but it was considered inaccessible because of the savage and hostile Cayapo and Chavante Indians. They "fiercely opposed any penetration into their domains," Oppenheim said.

Victor Oppenheim poses, pipe in mouth and gun on hip, in the state of Alagoas and Sergipe, Brazil, July 1933. Victor Oppenheim.

Portuguese, Brazilians and other foreign expeditioners had tried to penetrate the country. "All had failed, and most of them were lost." One of the last expeditions had been in 1925 when a British explorer, Colonel Percy H. Fawcett, had disappeared into the jungles. "Together with his son and another member of his party, he vanished, presumably on the way to the valley of Rio das Mortes. Nothing was ever heard from the three men and for years the riddle of Fawcett's expedition intrigued the press," Oppenheim said.

Despite the omens, Oppenheim ventured into the country. He usually traveled light, taking guns, ammunition, fishhooks, perhaps a tent and a few tools.

He expected to live off the land. "Alligator tails are excellent," he explained. The deadly piranja fish are "bony, but tasty." He also ate jaguars and roasted monkey. Armadillo, cleaned and baked in its shell was both appetizing and convenient. Peccary was a true jungle delicacy. Theoretically, it was easy to find food in the jungle. "But to feed six hungry men, undergoing daily exertion, totally on game was far from a simple matter," Oppenheim said. The noise usually drove the animals from the trail so that they had to be hunted down, often far from camp and after an exhausting day. The average game consisted, instead, of small animals, monkeys, rodents and wild turkeys.

Oppenheim and his men eventually came face to face with the headhunters. They were wearing rags of Fawcett's clothing, and one had Fawcett's suitcase. Other belongings were evident. Oppenheim managed to pacify the tribesmen. On other occasions, he was less lucky. He kept the shrunken head of one of his porters which he had recovered after a tangle with some Ecuadorian headhunters.

On a Shoestring

Between mid-December 1931 and mid-January 1932, P.T. Cox and A.H. Taitt were on restricted assignment in Iraq, close to the Kuwait border. "We went there without great hopes of finding anything very attractive," Cox wrote. "However, we first visited the Bahrah seepage and were encouraged to find what we believed to be an active gas seepage."

There were other encouraging signs, and they recommended further work in Kuwait. "I was sent back to Kuwait in February with a driller—John Green—and a makeshift portable drilling outfit capable of reaching only a few hundred feet. In those days of economic depression, things had to be done on a shoestring." They battled running sands and a lost drill bit, which they were still trying to fish out when they were recalled in April. Negotiations with the country's ruler took nearly two years, and the combined efforts of Anglo-Persian and Gulf, before a pact was finally signed.

Later that year, E.W. Shaw of the Iraq Petroleum Company and P.T. Cox were called on to survey Qatar, a small independent country which occupied a prominent peninsula in the western part of the Persian Gulf. The survey was conducted from January to April, 1933.

"We traveled from Abadan to Doha by a small steamer, taking a light truck and a Chevrolet car with us. As there was not enough water for us to get within seven miles of the shore, we had a serious problem in getting our transport to land. Apart from a gaudy Buick limousine which the Shaikh possessed, ours were the first motor vehicles ever to land in Qatar. At Doha, we were met by 'Haji' Williamson who was our interpreter and camp boss for the survey."

Shaw was already ill, which did not help matters. "Once ashore and having paid our respects to the Ruler, we started mapping the peninsula by prismatic compass and car mile-ometer traverses. I drove the car with Shaw beside me and two Bedouin guards in the back seat, singing camel songs most of the time with tempo adjusted to the speed of the car. Haji Williamson travelled with the truck carrying the camp kit, labour, and remainder of the guards...All we had to start with was a blown-up photostat of the Admiralty chart of the coastline. This was accurate as far as it went, but south of Bahrain, where the coast had been sketched in with a broken line, it was found to be some 40 miles out...Thus we made the map on which the concession was eventually based."

The Unsung Heroines

As the depression blanketed the states, some oil companies—particularly Atlantic Refining Company—continued to send exploration parties out of the country. "When major companies were firing geologists in the early phase of the depression, some Atlantic executives put on pressure to keep their exploration organization intact," Ed Owen explained. "Philadelphia decided that if they could find oil in Cuba it would be profitable even in the period of distress prices. They transferred geologists from Texas, Venezuela and Mexico for that program—everybody, I believe, who was not badly needed in his spot...The big campaign was dictated by considerations of company morale and 'public image.'"

Rolf Engleman and W.D. Chawner were in Cuba in 1931 and were joined by Aubrey Burns, a "tall, young diabetic" geophysicist, and Peter Kolm. Also in Cuba were Robert H. Palmer and Dorothy Palmer, who worked as a paleontological team.

Lewis Weeks had been in South America for Standard Oil of New Jersey since 1924, hopping from one end of the continent to the other. There had been trouble between Bolivia and Paraguay as early as 1928, but things had come to a head in 1932 when the Bolivian Army temporarily seized Jersey's producing wells. Politicians denounced the company for not lending them several million dollars to finance the Chaco War. In 1934, Weeks was named president of Standard Oil of Bolivia. Two years later, the company's properties were declared forfeit to the government. When a judge seemed inclined to favor the company's official appeal, the military dictatorship replaced the judge.

During the depression, many of the oil companies sent wives and children home. Men often found themselves sharing quarters, bachelor-fashion. "The life of an oil company prospecting geologist in my time involved not only physical hardship, but long and trying separations from loved ones," Weeks wrote. "The unsung heroines of the worldwide search for oil, which made our present economic prosperity possible, were the wives who stayed at home to keep the family and household together."

The Castor Oil Line

Walter S. Olson first met Thomas C. Wilson in November 1933, "at a jungle camp on the Venezuelan side of the border," according to Wilson. "We spent the next two years mapping the geology of the Barco Concession from camps reached only by mule trails or footpaths hacked from the virgin rain forest. It was hard labor, spiced on occasions by attacks from the Motilone Indians, but it was man-sized adventure."

Some geologists were lucky enough to be sent out of the United States for short missions rather than longer ones. Early in 1934, geophysicist H.B. Peacock was sent on a supervisory expedition to Venezuela where three field crews were operating for Caracas Petroleum. "My trip down by air was not without its 'moments,' " he wrote. "It was my first long plane trip, starting with a 'flying boat', a large four-engine plane, from Miami to Maricaibo. All went well until we reduced altitude for a landing at Cienfuegos, Cuba."

Peacock became violently airsick and did not recover entirely during his overnight stop at Kingston, Jamaica. "The hotel grounds were beautiful but I was in no condition to enjoy them," he wrote. Although the flight to Maricaibo was relatively uneventful, he was still ill and queasy from the first flight. "After another night's rest I was able to survive in a little better condition the flight to LaGuira, on a small amphibian, where we landed in the bay. A beautiful mountain drive took me to Caracas where I was forced to wait several days for a plane to Ciudad Bolivar, on the Orinoco River. There was no landing field at Caracas, and it was necessary to go to Maracay fifty miles or so away to take a plane for Ciudad Bolivar. This plane was on the famous 'castor oil' line, so-called because they used castor oil for engine lubrication. The fumes were terrible, and again I was sick most of the way. A short stop at San Fernando de Apure was not long enough for recovery and I was quite weak for several days after reaching Ciudad Bolivar.

"The field crews I was to visit were some hundred miles northwest and on the opposite side of the Orinoco from Ciudad Bolivar. Crossing the river on an old flat sailboat was quite an experience, and I was glad to be met by a car to take me to the field camps. In spite of almost daily rains, the roads and trails across the country were never very muddy as the surface was largely coarse sand."

The camps consisted chiefly of tents, but each crew had an office, a kitchen and sleeping quarters built by the natives, weaving mariche palm branches between poles to make roof and walls. "These were much cooler than tents and in spite of frequent rains they did not leak," Peacock wrote. "In these camps, I learned how to sleep in a hammock which was also woven out of mariche palm fiber. The 'trick' in sleeping in such a hammock is to lie neither lengthwise nor crosswise, but at an angle to both directions."

Peacock then set to work with geophysical equipment. "I must add that this area is much more difficult to work than I had expected," Peacock noted. "Contrary to reports, conditions are much more like those in the San Joaquin Valley of California than those in east Texas. The upper few hundred feet of sediments are coarser and more unconsolidated than even those of California, and these beds seem to dissipate the reflected energy so that we get either no reflections, or a profusion of minor ones."

The field crews, under the leadership of Bill Ransone, Morris Spencer and Bill McDermott, were having "rough going record-wise, and the unsympathetic 'help' " of two English geophysicists did not improve the matter. "Even in working along streams, conditions differed greatly among the three parties," he wrote. "After spending several days with each crew, I was suddenly recalled to Ciudad Bolivar for another round of conferences and discussions."

Whether it was the result of working conditions or economic circumstances, the company decided to shut down the work. Peacock left a few days later, but chose the long way out, down the Orinoco to Port of Spain, Trinidad, on a cattle boat which made several stops along the way to load cattle on the lower deck. Not until he reached Port of Spain did he venture back aboard another plane—this one a smaller but smoother one than the notorious castor-oil line craft.

Reading by Turner Gas Light

Canada's Turner Valley area near Calgary had been producing gas and a little oil since 1914, but it was a miserably small trickle. In the 1920s, 114 wells had been completed to extract condensate from the gas of the Mississippian formation. By 1932, condensate production in the valley had reached a rate of nearly 4,000 barrels a day, but the residue natural gas left over was simply flared. The giant flares were so intense that people in Calgary, 30 miles away, boasted about spending a summer's evening on their front porch, reading the newspapers by the light of the flares.

But the major oil field had not been touched. Robert A. Brown, an electrical engineer, George Melrose Bell, publisher of the *Calgary Albertan* newspaper, and J.W. Moyer, a Calgary lawyer, suspected that an oil column lay down-dip from the gas reservoir on the flank of the Turner Valley field.

Their theory, however, ran contrary to accepted geological thinking, which held that

On September 30, 1927, a snowstorm hit Pincher Creek, Alberta, Canada. The storm left roads at their worst. Thirty inches of snow on top of 15 inches of mud left geologists stranded. H.M. Kirk.

Manager C.H. Ford and a Cuban boy (Emilio Aleman) take the Atlantic trail into the interior of Loma Cunagua, Morón Province, Cuba, in February 1933. W.D. Chawner.

Camp staff and military guard all gather for the traditional Id-i-Noruz feast. N.L. Falcon.

Geologists take time out for a "bathing party" at Jask in 1933. Left to right: Dr. J.V. Harrison, R. McCall, D.A. Allisor?, the officer in command of the military escort and an Iranian government representative. N.L. Falcon.

In 1933 N.L. Falcon and several other geologists made a geological reconnaissance survey of the coastal Makian, Persian Baluchistan. The village of Jask is in the background in this picture. The geologists were not allowed through the country except with a military escort, but the officer in command of the escort confessed that the escort would not have been safe in Baluchistan had it not been for the geologists. N.L. Falcon.

only a large water table lay below the gas cap on the flank of the structure. Brown and Bell believed that the reservoir contained gas at the crest, followed by the oil column and then a water table. But to test their theory, they would have to drill the deepest well ever undertaken in Canada at that time. It was 1934, the middle of the depression, and money was not easy to come by. Investors had lost a lot of money already to less-than-honest promoters who had pulled too many fast deals. Besides, their geological theory was crazy.

Still, the men would not give up. They sold royalty interests in the proposed well in an effort to get the necessary capital. Seven times they stopped drilling when they ran out of money. Each time, Bell and Brown hit the streets, sold more royalties and came back with additional funds. Brown mortgaged everything he owned—house, insurance, car—to raise money for the well. Bell was also in debt from earlier unsuccessful wildcat ventures and his newspaper ventures.

Bell died in April, three months before the well blew in. The Turner Valley Royalties No. 1 blew in June 16, 1936, at 8,282 feet, with a roar heard throughout the valley. "The strike was made late Tuesday afternoon when accumulated gas pressure forced thousands of feet of heavy rotary drilling fluid up the hole and scattered crude oil over a wide area in the vicinity of the derrick," reported the *Calgary Herald* the next day. "It is the first time in Canada's history that anything approaching a crude oil gusher has been struck."

It was the opening of the largest oil field in what was then the British Empire. Eleven years passed before another significant oil field was found in Canada.

The Academic Side

C. Hewitt Dix had originally studied mathematics and expected to join the academic side of the fence. Oil companies realized his potential, but he was deaf to their courtship. He decided instead to complete his doctorate in mathematics at Rice University, in hopes of gaining an academic position. But formalities at the university delayed the official awarding of his degree for a year; and as a result, Duke University withdrew an offer of an assistant professorship.

In 1931, when he officially got his degree, it was "the worst possible time to be job-hunting," he told a reporter. "I think Rice felt sort of responsible for my situation, so they kept me on as an instructor in mathematics. However, my salary decreased every year because of the depression."

In 1934, Dix decided he could not afford another pay cut, so he took a job with Humble Oil Company. "Even though I was making more money, it was quite a blow to leave the academic world after so much preparation," he said. "I moped around a few days, but Humble was working on some interesting things and had lots of interesting people. After a few days, I began to get with it. In a couple of weeks, I was enjoying myself. After that, I never looked back."

Students on a geological field trip pose at Rock Canyon, Provo, Utah, between 1927–1930. Brigham Young University.

Burton Ashley and Sid Schafer talk with the Baroness von Blixen on the S.S. Springfonteur. The Baroness was better known in America by the name of Isak Dinesen, for her work, *Seven Gothic Tales* and *Out of Africa*. B.E. Ashley.

Field work frequently meant long days broken by evenings spent in rough camps far from the niceties of "civilization." A resourceful geologist, however, could make any camp feel like home.
B.E. Ashley.

Burton E. Ashley (left) and Sidney Schafer lunch after a day's traverse in Zambia in 1933—complete with teapot and tablecloth. B.E. Ashley.

A Bad Frame of Mind

When Glenn M. Ruby made his way to Argentina in the late 1930s, like many others, he sailed rather than face the rougher air travel. "My trip down from New York was uneventful, except for the Neptune party in which everyone who had not previously crossed the equator got initiated and ducked in the pool," he wrote. "I was the last one ducked because I did not submit meekly and, by using some of the tricks I had learned in my earlier wrestling career...my loyal son and some other neophytes [and I]...tossed the whole bunch in at once, breaking up the party."

Ruby was hired to reorganize Argentina's government oil company, Yacimientos Petroliferos Fiscales (Y.P.F.). "After a month I was able to read [Spanish] and now manage the language fairly well, but I will never be an accomplished linguist," he wrote. "I am the only American who has ever worked for this branch of the government, and they do not encourage the use of English. However, I am getting along well, so far, and they are adopting a lot of new ideas, mostly those which I have tried and know will work. Nevertheless, some...die-hards resent it and I have more than one personal enemy."

After tours along the Andes, in the central plains and the Pampas, Ruby struck out on a five-week trip to southern Patagonia, Tierra del Fuego and Cape Horn. Then he planned to head into the jungle country in northern Argentina to complete his preliminary survey. "I'm fed up with travelling, as I have never yet seen any place which did not look like I thought it would," he wrote. "The people, however, are different and they are the interesting things.

"In northern Patagonia [Argentina], the Territory of Neuquen, for example, has almost exactly the same climate as Arizona or Nevada. A little more wind than the parts nearest Los Angeles, perhaps, but it is a cold wind. The vegetation is almost the same and the modes of living should be the same, but they are not...Even the geologists have a very different system and very peculiar ideas. Sometimes I think that their mothers did not tell them where rocks come from."

Ruby was joined by three "very eminent" German geologists and three "equally eminent" Italian geologists, "all called 'Doctor' " and trained in the oil fields of Switzerland. "Each one, including myself, had a personal peon," he wrote. "This entourage required a sedan and three Ford pickups"—better known in Spanish as 'peek-oops'.

Ruby quickly tired of the Europeans' haughty attitude. "They were spending days to show me fossils and outcrops which I could have seen alone in a few hours, and explaining their very intricate hypotheses to me in an effort to be helpful." He was more put out with their attitude toward the natives, however. His temper was not helped by the bad cold that he developed and "which put me in a bad frame of mind."

When they tried to argue with Ruby "about something that I had known for so long that I had almost forgotten it, and knowing that they knew nothing whatever of petroleum geology, I blew up and began to run things."

The first change Ruby made was to get started before ten o'clock in the mornings, "and eat a cold lunch instead of taking three hours off in the middle of the day to prepare an asado. An asado...is a goat or sheep, skinned, dressed, split down the front and mashed out flat and skewered on iron rods and put over an open fire. Properly prepared, it is a very fine dish and is to the Argentine what baked beans are to the Cabots and the Lowells."

But Ruby felt the group did not have time for such niceties. "Anyway, as the doctors got more and more down on me, the peons rallied to my support. My peon, Vargas, was a half-breed Araucana Indian...about 25, and about as smart and alert an individual as I [had] ever seen anywhere. In fact, he was by far the most naturally intelligent one of the whole bunch...

"The second day out, he said to me, 'Doctor, if you do everything yourself, there will be nothing for me to do and I will be discharged.' I explained to him in my none-too-good Spanish that in North America the men were just as good as peons and were able to do things for themselves, and that if he would use his very good mechanical ability to see that the cars were always in proper order, that would be sufficient. I hoped to set an example to

In 1934, Professor J.P. Buwalda, dean of the school of geology at California Tech, led a spring vacation field trip. The well-dressed students included (left to right) Louis Hemniteer, Bob Sharp, Jack Judson, (?) Bakeman, W.D. Chawner, L.H. Evans, John Peter Buwalda, (unknown), Myron Hunt (president of the Rift Club), Maurice Donnelly, (unknown).

the foreign geologists and get them to do things for themselves and thus cut down on time and expense, but they gave me no notice and went on sitting on their folding camp chairs and sipping their yerba mate [Paraguayan tea] …while it took a peon for each one to run to the fire and keep him supplied with hot water."

Hospitality in the Andes

Ruby and the entourage reached the headwaters of the Picún Leufú and moved into the foothills of the Andes. They made their way south to an out-of-the-way estancia which had been bypassed when the roads had been changed. "It took most of the day to reach the estancia, as the things called roads in this country were designed for the enormous high-wheeled carts, [called] carros, still very much in use," he wrote, "and the new type of automobile is much too low-slung to go far without pick and shovel."

When they arrived at the estancia, Ruby found the men out on roundup, and the windmill which pumped water was broken down. Only the girls were there, working on their costumes for an upcoming fiesta. Ruby put his peons to pumping in relays in order to fill the water tank.

"The doctors did not turn a hand at anything, not even to get their own baggage out of the pick-up." Ruby was growing shorter tempered all the time. "The hotel consisted of a series of rooms in a long, adobe house with a veranda paved with ladrillos, the large, flat, home-made, slightly burned bricks. The cold, gloomy rooms were also paved with the same ladrillos. While the doctors stood around waiting for someone to put their voluminous baggage in their rooms and fix their afternoon mate, I fished out my solitary suitcase and sent Vargas to pick up some dry twigs and make a fire in the kitchen, so the girls could heat a kettle of water. By the time I was settled in my room and had my mate equipment, including the bombilla ["little pump"], the small silver tube through which the drink is sucked, Vargas was ready with some hot water and prepared my mate. I wrapped up in my heavy wool poncho, which I had purchased a few days before as protection against the cold winds, and settled myself in an armchair on the veranda. This made the doctors all the more anxious to have theirs, and one of them went to order the peons off the pump in order to unload their luggage, etc. from the pick-ups—an order which I very promptly and firmly countermanded."

Then the doctors decided to employ the women to unload. Obediently the native women followed them out to the pickups. They were about to begin unloading the stuff when Ruby decided he had had enough. "I told them not to bother if they had other work to do," Ruby wrote. "One girl, however, said she was not going to the baile [dance] and had nothing else to do. She didn't weight [sic] over 100 pounds, but those foreign bums stood idly by while she got on the truck and began to haul out baggage. The final straw with

"Howdy" Kirk sports a panama hat as he starts across the Euphrates River in Turkey in 1934. "I think the boat was one they used to send animals to land on Mount Ararat," Kirk wrote. H.M. Kirk.

me was when she lifted over a big, heavy duffle bag to have one of the doctors set it on the ground and he refused to even touch it. I ordered her off the truck and told her to go back to the veranda and sit in my chair. Then I told the bunch of wise men to unload their stuff, and do it damn quick or the geological department of the Y.P.F. would, from that moment on, be deprived of their services."

The doctors were impressed. "They immediately changed their whole attitude and not only went to work, but seemed to like it. It must be that they like dictators, for those same men are now my very best friends and treat me with the utmost respect. However, I still let them click their heels when they shake hands and do not tell them to sit down when they come to my office.

"About dark, old Alonzo and a bunch of his gauchos rode in and Vargas, whom the old Don knew and apparently liked very much, brought him to my room. He is the biggest man I've seen in Argentina. At least five feet and nineteen inches tall, and weighs a good 20 stone [280 pounds]...Vargas had told the old fellow about the baggage incident and his old black eyes sparkled like the spokes in a brand new squirrel cage. I received the most cordial welcome to the place and felt the most sincere hospitality of my life."

The group sat down to supper about ten o'clock, which was the usual hour for the Argentines. "The Doctors were in the best humor since we had started on the trip," Ruby wrote. "I know now that it was because they had found a boss who did not treat them as equals."

Supper, or ceña, Ruby explained, was an institution in Argentina and made up for the scanty breakfasts.

"Any one course would be a whole meal in the States, but it is so well served, and so well washed down with wine, that you can eat almost indefinitely.

"First we had fiambres, an assortment of cold and pickled meats. This was well washed down with very good red wine. Then while the girls cleared out the plates, we talked and drank some more red wine and munched criollo bread. This bread is shaped like a French roll and has the consistency of cork. I think it accounts for the very fine teeth the natives have.

"Next came the soup. I have never eaten, or swished a spoonful of, unappetizing soup in the Argentine...After the soup was more red wine. Good wine is very cheap in this country and they consume a greater gallonage of wine than gasoline.

"Next was the tortilla. It is not made from cornmeal as in Mexico, but is a sort of omelette...More 'pan criollo' [pan = bread] to gnaw on with this course and to wash down with the constantly full glass of wine, while the table got its customary clearing of plate and cutlery. "Then came fish...This fish was a sort of trout, it looked like our German Brown, and with it we had some exceptionally good white wine in what looked to me like real crystal glasses...

"The main course was asado de bife. It was really a prime rib roast weighing about 40

**Victor Oppenheim traveled by native boat on an expedition to the Upper Juriva River in the territory of Acre, western Brazil on the border with Peru in 1935–36.
V. Oppenheim.**

In 1934, Victor Oppenheim (center) was consulting geologist to the government of Brazil. Mark C. Malamphy was a geophysicist working in the Paraná basin. V. Oppenheim/AAPG.

pounds, and had been roasting over the coals in a small lean-to next to the kitchen ever since the peons had returned from their trip for wood. They always use an open fire for the asado and it must be the coals from a certain wood or brush. It closely resembles the Iron Wood which is so rapidly disappearing from the Imperial Valley.

"The roast was so heavy that the girl nearly fell over my chair when she set it down for the Don to carve. All other courses are served, individually, from the platter, but the Head of the table must carve the roast...

"There was no carving knife, but Alonzo merely reached back with his right hand as though he were going to scratch his back...and pulled from its sheath, or vaina, a very sharp, silver-handled knife with which he began slicing the asado but not until after he had carefully investigated the edge with a saliva-moistened thumb.

"The knives are a necessary part of the properly dressed gaucho's apparel. Everyone in the country carries them, stuck through their faja, or wide, hand-woven, colored, woolen belt. They are stuck through the belt at an angle so the handle is near at hand by a quick, backward grab. In camp, they are the sole eating instrument and I must say I have found the system quite convenient. You cut off a chunk of hot asado with the knife; then you hold

(Opposite page)
In 1936, AAPG was headquartered in the Wright Building in downtown Tulsa. Pictures are of "Misses Robertson, Cummings and Whalen." AAPG.

Jack Huner (left), Fisk (right). The drill stem length was 26 feet. W.D. Chawner.

Dr. Harold N. Fisk (left) and Jack Huner augur hole for formation samples in La Salle Parish, Louisiana, for the Louisiana State Geological Survey in April 1936. W.D. Chawner.

The Tropical Oil Company camp, Barrancabermeja, Colombia, in 1935 at which Robert E. King worked. The camp included well-built tropical homes with paved porches and outdoor furniture for evening relaxation. R.E. King.

the asado in one hand the...knife in the other and bite into the meat; then you run the knife along your teeth and sever the bite from the chunk. Until I got over the fear of cutting off my nose, when I cut upwards, and my chin, when I cut downwards, I can truthfully say that I cut off some mighty big mouthfuls.

"After our plates had been removed again, it was necessary to submit to the last course which was a native cheese, [or] queso, and membrillo. The latter is a sort of jam or jelly, made from quinces, and is spread over the cheese. It is quite good even if you don't need it. Then came the substance the Argentines call coffee...It is so thick you almost have to jab your spoon into it."

After the coffee had been served, the Don announced that he was tired. "It was then midnight for we had been eating for the customary two hours. He got up from the table and, as he rose, he told me he had arranged for a hot brick (ladrillo) to put at my feet as it was the proper thing to do. A man with a cold should eat much and sleep warm."

Ruby felt considerably better the next morning and got up late since it was a holiday, and he knew that no work would be done by the peons. "The men and women on the estancia had on their best outfits and several of the neighbors had come in to join in the fiesta.

"I ate my breakfast of pan criollo, membrillo and coffee alone as everyone else had finished and gone out to watch the natives...I was sitting in front of my room, on the veranda, when Alonzo brought up a chair and sat down with his mate. He asked me how I felt and if I had had my mate yet. He called Vargas, who took the Don's mate to the kitchen and filled it with hot water and when he came back, the Don handed it to me. I sipped on it until the water was gone, when Vargas again filled it and gave it to Alonzo. When it was again emptied, it was returned to me. This is the real mark of Criollo hospitality; when the host asks you to share his mate ...your name is mud if you even think of wiping off the mouth-piece of the bombilla, or sipper, when it is passed over to you."

There were no regular meals that day. "Everyone ate from a big table loaded down with all kinds of food and wine, and the supper was not served until daybreak. With the rest and the diversion, my cold was much better." Ruby and his party stayed for the remaining holiday festivities, then headed out for the more unsettled country. From that point, he had little trouble with the 'Doctors.'

Two Hawks and a Greyhound

In the mid-1930s, activity in the Middle East was heating up. Linn Farish and Fran Reeves were surveying Iran for the American Oil Company. In 1936, T.F. Williamson and D. Glynn Jones were hired to make a comprehensive examination of the Trucial States for the Iraq Petroleum Company.

"We landed from the Gulf mail steamer at Dubai, after a day and night lying to and

Michel T. Halbouty (front) and Glenn H. McCarthy (left of man bent forward) watch at the opening of the discovery well in West Beaumont, Texas, field in 1936. M.T. Halbouty.

**John Muir, chief geologist for Island Exploration Co., was well known for his geological work in Tampico, Mexico. He headed for Port Morasby, Papua, via Sydney, Australia, in December 1936 aboard the Van Rees. He died of a heart attack in Sydney a year later.
W.D. Chawner.**

waiting until one of the most violent storms of thunder and rain we have ever experienced had blown itself out," Williamson recorded. "During the first few months in Dubai we were advised never to leave the house without an armed escort, as we were the first Europeans to live in the town and it was considered desirable to accustom the local populace gradually to our presence.

"We made a number of trips inland from Dubai to cover most of the Sheikh's territory, and on these we were usually accompanied by the Sheikh himself and a number of his followers, who welcomed the opportunity for a hunting trip. With our own two cars, we had the Sheikh's and at least one other filled with his men, six or eight to a car, each with his saddle bag of belongings and his rifle. Room had also to be found for a few cooking pots and one or two hawks, while on one occasion a large saluki (greyhound) was also squeezed in. Our routes followed camel trails, the only tracks of any kind which existed, and as most of the country was sand the cars were very often stuck."

The group made their way to Abu Dhabi, arriving just before a four-day holiday. "As no work could be contemplated until after the holiday, our forenoons were usually spent in rather solemn visits to important people, each beginning with a round of handshaking. The Trucial Coast greeting is a light ceremonious touch of fingers only. Innumerable cups of coffee were drunk on these visits and "halwa," a locally manufactured form of Turkish Delight, had to be tasted."

Soon afterwards, they drove to the Buraimi Oasis, where they saw what Williamson termed "the only interesting bit of geology" during the whole season.

From there, Williamson went on to survey Qatar. At that time, he employed one of the best local guides he ever encountered, Sheikh Mansour. "Some have been very good indeed," Williamson noted, "even when using what was to them an unfamiliar mode of transport, the motor car; others were utterly confused by the speed at which cars covered the ground; a few seemed to depend on the sun for their orientation and were helpless on a cloudy day. But Sheikh Mansour seemed to be quite unaffected by any outside influences. He appeared to have a complete mental picture of the whole of Qatar and had a quite uncanny capacity for knowing exactly where he was under all conditions—in clear weather, in fog, or even in darkness....Mansour had a quick and clear appreciation of our ideas and requirements and, as his standing was high with all classes of the local inhabitants, his advice on many occasions prevented disputes and smoothed over difficulties." (Although Mansour later went blind, he conducted Noval E. Baker over the featureless peninsula in 1946. Mansour's nephew acted as seeing eye. On that occasion, Mansour managed to locate new outcrops of the famous Alveolina bed and seemed to know every bump on the camel trails and even the bushes.)

The rider on the left may have been Kaslite (a German citizen); on the right is Bela Hubbard. Photo by C. Chaveri, Cairo, Egypt. B. Hubbard.

167

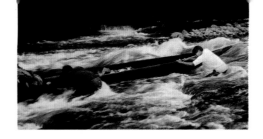

Fighting the rapids was just one more jungle hazard when Victor Oppenheim explored the Río Pastaza in eastern Ecuador in 1938.   V. Oppenheim.

(Opposite page)
Victor Oppenheim's team build a raft to explore the Río Morona in the Oriente of Ecuador-Peru, in 1938.   V. Oppenheim.

Victor Oppenheim (left), then chief of exploration for Royal Dutch Shell, descended the Río Macuma, Ecuador, with assistant Julio Granja.
V. Oppenheim.

Jungle rapids were the undoing of many an expedition. Victor Oppenheim's crew cross the Rio Pania rapids with trepidity.
V. Oppenheim.

Mule trains were staple transportation in the South American jungles. An expedition descends from the Andes to the Rio Napo valley, in the Oriente of Eastern Ecaudor, in 1938. V. Oppenheim.

Full speed ahead by paddle power on the Rio Aguarico, Ecuador, in 1938. Some of the more spectacular wells in South America were drilled along the Aguarico. V. Oppenheim.

Victor Oppenheim and crew pushed up the rapids of the Rio San Miguel de Sucumbios in Ecuador, near the border with Colombia, in 1938. V. Oppenheim.

Oppenheim and his assistant, Walter Ziugg (right) camp in a Jivaro Indian clearing in Eastern Ecuador in 1939. V. Oppenheim.

In 1939, Chief Jimbiquique of the Jivaro (headhunter) Indians of Ecuador posed with his five wives, several babies, two grown sons, two muskets, one dog and Victor Oppenheim.   V. Oppenheim.

Cognac on the Campfire

H.M. "Howdie" Kirk was working for the Turkish government in 1937, when American Oil cabled him from New York offering him a position as senior field geologist in Afghanistan. His job would be to survey the Registran and Gaud-i-Zirreh deserts. "I returned to the states (by boat), attended to personal matters, and in three weeks time I was on a freighter with my gear sailing for Karachi, Pakistan (then India)," Kirk wrote.

"While in New York, I was advised that I should live off the country. I can come as near as anyone living off the country and have done so at times. I also know the time it requires to attempt to do so. So with this background, I proceeded to stock a year's supply of groceries and had same shipped with other supplies. When working for a company without foreign background, it is better to be your own purchasing agent.

"On the trip with me to Afghanistan were…Henry Rea and Ernest Fox. Rea and myself were on an oil assignment and Fox on a mineral search. On our arrival in Karachi, we cleared customs and shipped our Marmon Herrington station wagons and our equipment via rail to Peshawar."

From there, Kirk drove overland via the Khyber Pass to Kabul, where they stopped for a few days. Then they proceeded to northern Afghanistan via Shibar Pass (9,800 feet) in the Hindu Kush Mountains.

"It was decided Rea would work the western area north of Hindu Kush Mountains, and I the eastern. We chose the city of Mazar-i-Sharif as the division point, and left our supplies in a storeroom."

They agreed to meet back at that point on Thanksgiving Day, but Rea had an acute appendicitis attack a week after they parted company and had to return to Kabul and Peshawar. Kirk did not see him again until late March.

"There was a passable fair weather road between the cities of Mazar-i-Sharif and Khanabad and a few of the villages," Kirk wrote. "But these roads for the most part did not cross any areas of geological interest. Most of my field work was done on foot or horse back. Some of the time I would take a light pack horse outfit for a week's circuit trip from a car; then move the car up the road twenty or thirty miles and repeat the process."

At Khanabad, Kirk paid his respects to the Governor, who tendered him a dinner at the Government building. "Table clothes were spread on the floor of the main hall of the building, and a great variety of dishes were offered. Every time I had eaten enough, more food was offered. My appetite could not keep up with all the food offered.

"The Governor insisted I have guards accompany me on my work and he assigned two fine, heavily armed men to accompany me. The company had hired a chauffeur for me and he drove from Kabul to the north, but as he lacked any courtesies of the road to other travelers, I got rid of him. In addition, I had an interpreter, Mohammad Asif, a man of fine character; but entirely inexperienced in camp life. Due to the many different languages or

The steep sand dunes of the Guad-i-Zirreh desert were soft footing even for a camel. "We had to tamp steps for the camels to climb over on," he wrote. "The white ground is alkaline clay soil. On this trip, the government officials insisted I take soldiers with me, so I had a large crew." H.M. Kirk.

dialects in the country it was sometimes necessary to have an interpreter for the interpreter. I never was certain after a message had passed over three or more tongues, whether I got the intended picture."

There were enough outcrops and topography to make reasonable assumptions as to the area's structural patterns, and Kirk outlined six anticlinal features which he recommended for detail mapping. He was surprised later to read a cable in the Kabul office, which stated "Kirk mapped six excellent drillable structures."

"One day I was working waist deep in mucky water and with a stick was agitating the bottom of a still-water stream. Large bubbles the size of an inverted salad bowl would form and I would ignite same with a match. This caused me considerable prestige with the natives accompanying me. This was further augmented the same evening when I secretively filled my mouth with cheap cognac, which ignited when I spit on the camp fire. I say cheap cognac, as I would not have wasted a good brand."

Pack animals in a caravan could usually travel about 26 miles a day. "The best camel I ever had could make 40 miles a day," Howdie Kirk wrote. "It threw me off three times before I qualified in its opinion. I am not holding the toggle rein properly, so don't show this to any desert traveller."

In the fall of 1937, Kirk worked in the north and used pack horses for trips off existing roads. F. G. Clapp, who was in charge of exploration in Iran and Afghanistan, spent a few days with Kirk in the field in December. "The long days and cool weather caused hearty appetites. We would stop for the night and while the others got the tents up, I could have a meal prepared on a couple of primus stoves, thanks to the larder I had brought from New York. Mr. Clapp commented on the fine food, and I didn't mention that he had suggested that one live off the country.

"I left the northern area December 22, 1937, with icy roads over the Hindu Kush Mountains [and] with temperatures minus 22 degrees F...Due to three feet of snowfall, I was unable to leave [Kabul] until January 10 to drive to the south."

Flour for the Camels

Kirk's next assignment was southern Afghanistan, which was almost entirely desert. Nothing was known of the area geologically at that time.

"I drove a car south from Girishk on the west side of the Helmund River, following a well traveled camel trail some 20 kilometers," Kirk wrote. "A passing camel caravan gave M. Asif a ride to the village of Darweshan, where he obtained pack and saddle camels to come and move our gear and ourselves to the village. The car was left parked at the river. It was a couple of hours' trip to the village, and it was my first ride on a camel, and one felt like they were riding high."

At Darweshan, Sheik Hakim, the mayor of the village, refused to see Kirk until the mayor dressed in his official attire. "Thru him [I] made the arrangements to rent camels and drivers for the trip, and also got some information as to the water holes and route."

Kirk guessed that pack camels would make about 26 miles a day. "I planned to take a southeasterly route to the Baluchistan border, then parallel the border to the west for a couple of days, then swing north and circle back to my starting point. I made an estimate of 15 to 18 days time would be needed."

Kirk took along an interpreter, five pack camels and two riding camels. A cook was added while in Kabul. "In the afternoon I checked on the animals...For the camels, they had only one sack of bran to supplement their grazing on the camel brush. I knew nothing of camels, but told M. Asif to get two more sacks of bran. This could not be found in the village, but we could obtain flour (about 200 pounds were available)."

When Kirk told the interpreter to buy it, the interpreter asked why. Kirk knew the man would not understand his concern for the animals; instead he told him that he could sell the flour when he reached the southern part of the journey. "He thought that was a good idea," and the flour was purchased.

"M. Asif and the cook rode one camel and I the other. I carried a bedroll and emergency

In 1928, H.M. "Howdie" Kirk mapped northern Afghanistan. "I worked with pack horses away from the roads," he wrote. Face masks were often necessary to keep the sand from the lungs. H.M. Kirk.

It may have been desert country, but at Chahar Burjak, Afghanistan, in March 1938, Howdie Kirk had to ford the Helmand River in flood season.
H.M. Kirk.

food rations on my camel. I had been somewhat dubious in regard to riding camels, but my fears were ungrounded, as after one gets used to their swaying motion, they are easier riding than a horse for long trips."

Kirk found the gravel-strewn plain and the sand dunes monotonous at first. "As one progresses, [his] eyes become more observant," he wrote. "On the fourth morning, M. Asif advised me that the camel drivers wanted to turn back. I asked why, as I had explained before leaving it would be over two weeks trip. They said that they had been told that the trip would be too rough for me and I would turn back after two days, but that I hadn't done so. Also they needed more bran for their camels. I told them I would let them use my flour, and they agreed to continue. To feed the camels bran or flour, it was necessary to mix it with water and make a dough ball about the size of a soft ball. The camel driver would then open the camel's mouth and push the ball down its throat. We therefore saved all the water, and did not use soap for washing dishes or ourselves. The camel drivers thought that anyone who would give his flour to camels was an 'alright guy' and for that they would continue the trip."

There were long spaces between watering holes; and at one point, they traveled three and a half days before they came to a source of fresh water. "The water we carried tasted very flat after sloshing around in the water cans for three days," but they were glad to have it. "The weather was cool in the daytime with frequent dust storms, and cold at night. I wore a knee length sheepskin coat. The camel drivers all wore heavy felt coats, but M. Asif and Shasawa, the cook, were not so well protected and sometimes had to wrap a blanket around themselves. The biggest discomfort was the sand. It seemed to be in everything and especially in the food. I had not shaved since leaving Kabul and my beard protected my face from the blasting sand. Most of the time I traveled ahead of the pack animals and one evening reached our stopping point by the water hole. When the pack train didn't arrive I made a meal from the emergency rations and rolled out my sleeping bag. When I awoke in the morning my beard was covered by hoar frost, and I looked the part of the old man of the desert. The pack train had arrived and a hot breakfast was greatly appreciated."

In some areas of the desert, there was not even a water hole. Caravan members fill up their water tanks before starting on a three-day stretch where there was no water in the Dasht-i-Margo. H.M. Kirk.

The trip had begun to affect Kirk's health. "On our tenth day out I was passing mucus-bloody stools, so I took the day off, treating myself with stovarsol, a purgative, and confining my food to rice broth. I felt that the trouble came from eating sand with all my food. My day of rest [revived] me and also the camels, which were tired. I tried to hire new camels at a nearby village, but none were available."

They encountered only one rough wind storm. "This was at our backs so we struggled along to some protected brush near what had once been a lagoon. I arrived first and walked back to meet the camel drivers. Their faces were a mass of fine sand, and they were most grateful for me to lead them to shelter. We got a tent up and the reliable primus going, and tea for all made us feel better."

On the 21st day, they arrived back at Darweshan. Everyone was exhausted but elated, and "a goat was purchased and barbecued for all the crew." Kirk was the first foreigner to have made the trip across the desert.

The village chief, or Hakim, of Darweshan, posed for Howdie Kirk just before he began his first camel trip in the Dasht-i-Margo, the Desert of Death. H.M. Kirk.

Lost in a Sea of Sand

Kirk then prepared to cover the Gaud-i-Zirreh desert. "This time it was a larger troop as it was insisted that I take two soldiers and a guide with me," Kirk recalled. "It was about a ten-camel caravan. "We forded the Helmand River near Chahah Burjak and proceeded southward. We climbed the river bluff and were soon in the desert, which gave way to a series of playas. The area is reported to be devoid of any rainfall; however, windstorms carry water from the inland lake of Daryachen ye Sistan in Iran near the Afghanistan border and deposits some water in these playas."

For two days, they traveled through the playas and into dunal sand. "The wind was always with us. At lunch, I would hunker down on the lee side of my squatted camel to eat my sandwich. The camel would swing its head around for a serving. Its breath had the odor of a sewer and I named the camel 'Halitosis.' Our journey continued to the southwestern corner of Afghanistan, Malek Siah, where there is a boundary marker between Afghanistan, Iran and Baluchistan...I mapped a fault here, which was also mapped by Frank Reeves who was working in Iran, and by F.G. Clapp in Baluchistan."

When they started back, the guide insisted that he knew the trail, and Kirk turned the lead to him. "Within a half days travel on the return trip, we entered an endless sea of sand, and with some wind always with us. I usually traveled in the van of our caravan and left the route to the guide. One overcast day, I became doubtful if the guide really had any landmark to follow, and so I fell back about a quarter of a mile in order to make a compass check on the bearings he was following."

Kirk realized that the course was taking them into Iran, and he took over. "I changed course to east-northeast, and proceeded. About 4 p.m. that evening the sky suddenly blackened and a storm was nearly upon us, when I discovered that M. Asif and one of the soldiers were missing. I was told they had stopped to smoke the hubble bubble pipe. Visibility was nearly zero, so I had the other soldier fire his gun in the air. Fortunately they heard the shot and were able to find us. We got on the lee side of a large dunal hill and were able to get one tent up and anchored before the storm hit...I am sure if we had not had the protection of the hill, it would have swept our tent away. We were able to get the primus stove going and made tea with barley bran for all the hands. By morning the storm had passed and the quietness seemed unreal."

When they finally spotted the Helmand River, it was in flood. "I was told it would be six weeks before we could cross," Kirk wrote. Not to be outdone, he journeyed upstream about six miles to where the river was about a mile in width. "We had swimmers and horsemen come from Chahh Burjak. With these swimmers and outriders we put our caravan into the river and the camels floated and waded across the river."

Kirk moved on to look at western Afghanistan. "I had thought that I was finished with camels, but the car broke down, and while waiting some ten days for a part to come from Kabul, I again took to a camel pack train. About this time, I was joined by G.L. Postle, a geologist who arrived to help in mapping and to run a plane table." Shortly after Postle's arrival, however, the company decided Afghanistan was not a hot petroleum prospect, and they abandoned the project.

An ancient desert city abandoned when the Helmund River changed its course. H.M. Kirk.

Jan Baak, a Dutch geologist, listened as one native interpreted for the nomadic Monoways that he and Floyd Ayers met in the New Guinea jungle. "They were the only people we met during our survey for Standard Oil of California," wrote Floyd Ayers. "The native leader shaved his head about half way in the forefront, and a row of split boar teeth was woven into a band and tied at the back. On the extreme rear of the head, a large group of black Cassowary feathers stood out as a trademark of identification for this group. A chain of seeds was woven and placed over the cluster of feathers to hold them in place and then tied under the chin. The remaining part of his array was composed of various sized seeds woven into a wide belt. To protect the vital organs, the epidermis of certain plants or trees was used. Red cloth—one of their trading items—was also used when available. The horizontal stick through the lower part of the septum of the nostrils apparently established their manhood." F. Ayers.

Flinging the Fantastic Toe

In 1934, the Dutch East Indian government had announced a "Policy of Prudence" in relation to new oil rights. Borneo, Sumatra and Java were already producing oil at the time. But the land which the Netherlands Pacific Petroleum Maatschappij (N.P.P.M.) drew as a result of the new policy was a tract in central Sumatra which seemed fairly unpromising.

For five years, oil men and geologists struggled through the terrain without success. Results were so poor that some claimed that N.P.P.M. really stood for "Non-Producing Petroleum Maatschappij."

In 1937, while N.P.P.M. was struggling with Sumatra, William Argabrite was in New Guinea. Argabrite's team was assigned to do reconnaissance work in the tropical jungle, where conditions were still primitive. Some groups of natives with whom the geologists

A typical tent camp on the trail west of Kiunga base camp on the Fly River in the Australian territory of Papua, island of New Guinea, January 1938. W.D. Chawner was working for Island Exploration Company at the time and later worked for its successor, the Australasian Petroleum Co.(APC). W.D. Chawner.

Jan Baak, working for Standard Oil of California, poses on a thick iron-stained limestone formation in northern New Guinea. These formations were an important reservoir for petroleum. F. Ayers.

W.D. Chawner navigated the Fly River in Papua with a double canoe equipped with an outboard motor. W.D. Chawner.

came in contact had never seen metal objects. "An axe or even a kitchen paring knife was an object of wonder," explained Clarence Ross. The geologists progressed from camp to camp, "inviting the leaders from the next camp area to a big entertainment, a 'sing sing', presenting them with a new axe, and urging them to plant an extra crop of sweet potatoes, thus arranging for food supplies in advance."

A geologist in Madang, New Guinea, had more difficulty dealing with the remote conditions than did Argabrite's group. He wrote to an old friend in 1938, "Cleaning up a few old files and ran across your letter of July 16 [four months old]. Not a year old yet, but nevertheless I'll write. Really have been too damned busy to spend much time at a typewriter except on reports and Work Orders. The company has done quite well by me on my suggestions and requests, and we now have a force of 22 men here to take care of the work. The 269 native laborers don't count, of course. I find too that there is a lot besides telling people what to do to be done. Some of them you could tell a half dozen times and still have to go out and show them. Have had two of the youngsters from Australia quit and go back. One had a sweet young thing come up here so that he could marry her, and she took him back. The other had his wife come up for a few weeks, which stretched into six months and he had to go back as there was no chance for her [to stay]...You may have known Argabrite around Eastland in 1918. He is out here in charge of one of the parties. Getting on pretty well for an old man, too. Speaking of age, the fact that the women of this country are terrable [sic] things to look at don't [sic] mean a thing to me. No holes burned in the blankets by them. The only enjoyment I seem to be able to get with the fairer sex these days is companionship in drinking and flinging the fantastic toe about a bit...Had a number of good parties here during the Xmas and New Year week. Even a race meet with less than seventy five in attendance. I made a few dollars above expenses, too, because I didn't know enough about horses to bet intelligently.

"E-- and M-- must have had a good party a few months ago, for they wrote me a letter, each writing a paragraph in which they had little to say except that they had to stop and take the drink Earl was trying to get rid of. Beer comes in big bottles over here and there are four dozen in a case. Even so a party here usually needs two cases plus a few bottles of gin and some whisky. I don't think that these people know what is good to drink for they don't take to the whisky soda as I do. I get my share, usually, whether I can take it home or not. Have not been down too many times as yet. Getting better all the time, too. Maybe that is because I have not been getting on as many benders as I did for a while here. Had to call a halt or get someone else to do my work for me. Too poor for the latter.

"Someone wrote me that it was rumored that Burton Hartley was now with you. If so, tell the long hungry whatnot to write to me...I don't know yet just what this job is going

Suki tribesmen of the Fly River approach geologists' ship as they head upriver in January 1937. "They have marvelous balance," wrote W.D. Chawner.
W.D. Chawner.

W.D. Chawner (left) and K. Washington Gray sail leisurely down the Vailala River, Papua, as they tour the exploration work in September 1939. W.D. Chawner.

Eugene O'Driscoll makes a plane-table survey on the Fly River, Western Papua, August 1937, for the Island Exploration Co.   W.D. Chawner.

Natives built houses for their Island Exploration Co. employers on the island of Daru, Papua, in 1939.   W.D. Chawner.

(Right)
The Island Exploration Co.'s supply ship, Maira, was a unique vessel which made its way 485 miles up from the mouth of the Fly River in western Papua to the IEC base camp, Kiunga. The photo was taken December 1937. W.D. Chawner.

to work into, but there are possibilities and I will have to find things out for myself here as there has never been any work done in this country. The New Guinea government never had a survey and there is absolutely nothing to check back on. One thinks he has something good, then he finds out that he is either a lot higher in the section than he thought or that there is only a thin section down to the basement rocks. I do think, tho, that I have an area which will prove to be well worthwhile. There is the possibility, too, that some of these outlying islands will prove to be attractive. I want to get out among them this summer. I have more to do right now than I can get done but hope to get caught up by June or July, and take a month or two off, then, for a trip out around. Maybe I'll see something out there that will look better than the women of this place. A year and a half is too long...

"Hope that the recession, depression or cessession [sic] has not put too much of a damper on the oil business. Would like to see Earl and some of the others make a little good money before it is too late. It takes so damned much to carry a man thru one of those lulls, that he has to make a lot while the going is good. I never did quite come out even. It always cost me just as much to live without working as it did to work, and when there was no profit coming in it was too bad. Think that I was cut out to work for someone else, anyway. The salary is fairly sure usually."

Floyd M. Ayers was assigned to New Guinea by Standard Vacuum Sumatra's (N.K.P.M.)

office in late September 1937. He worked the geology of the Aika River and its tributary influents for the next three months. He and Cass Baumann worked together during this survey, "perhaps the first geologists to actually map this area," Ayers wrote. "It was a kind of survey in which a hundred-meter rattan tape secured from the jungle was measured at the time the trail cutter made it. A stake was thrust into the ground as he turned around to shout or yell, so that the other geologist could take a bearing on the voice ahead. This means of carrying a line from one river or influent across to the next one was practical for reconnaissance type of mapping.

"It rained some twenty days during a month during the time we were doing our survey. Rain usually started around 11 o'clock and kept up until 5 o'clock. These rains are not like a storm. There is no thunder or lightning, it just pours and pours."

The real problem in the jungle was the leech population. "There is no wind in the jungle, so it must be heavy rain that causes the low bush to be full of leeches and the floor of the jungle to be covered with the bloodsuckers," Ayers wrote. "Several times upon retiring for the night when the boot [was] removed I found my sock stuck to the bottom of the boot...A leech had been there and left. I never found one there, nor did I even feel its bite."

Near the end of the hard rainy days, on their return to camp, Baumann said he had a leech in his eye. "We had two dry matches in a tin so they were used as forceps [sic] but with little success. When I had the leech stretched out to the point of release, I thought he would now come from the eye. He only went further in the eye, as they are like a piece of rubber or elastic." About two hours later, the leech dropped out of its own accord.

Ayers found one inside his left nostril. When he blew his nose, it came out. "I did not feel it being there nor was any damage done to me." On one occasion, he noticed that the dark ground was covered with leeches. He stepped forward with one foot and drew it up, leaving a depression on the ground. Immediately, the leeches congregated towards the footprint. "We took to our camp without hesitation," Ayers said.

Wild Men of the Jungle

West of New Guinea lay Borneo, another unexplored island inhabited by headhunters, snakes and other generally unfriendly creatures. Shell Oil was ready to wager there was petroleum potential in the tropical jungle, and they hired three young geologists to investigate.

They made their way up the coast to the mouth of a river, stopping at a friendly Malay village. The head man agreed to take the group into the interior as far as the 'mad water' (rapids and waterfall). But they would not go into the country of the Dyaks, who were notorious headhunters.

Bananas were available by the bunch when W.D. "Bill" Chawner stopped at the Paku village "rest house" on the Vailala River, Papua, July 1939. W.D. Chawner.

The Australasian Petroleum's Paku base camp for geological parties on the Vailala River, Papua, might have passed for a native village. W.D. Chawner.

The Malays were true to their word; but when they arrived at the destination, an assistant chief and three young men armed with cane blowguns, poisonous darts and one canoe remained with them. The geologists set out through the jungle, planning to be back by nightfall, but they misjudged the distance and were forced to camp in the jungle. All night long, strange figures moved around them, but there was no attack. In the morning as the geologists shouldered their packs, nude painted figures with bone ornaments in their noses and hair approached. They were armed with blowguns, spears and jungle knives. There was a tense moment as the geologists tried to decide what to do. One of the men was adept at sleight-of-hand tricks; and while the groups were trying to communicate, he walked up to one native, waved an empty hand in front of his face, made a pass toward the man's mouth and plucked a white egg from between the man's lips. He explained that he had taken a sickness out of the man and saved his life. The excited headhunters led the geologists back to their chief's village, where the geologist performed the trick again. Then he pulled a huge bouquet of artificial flowers from under the chief's left arm.

At that point, the magic went awry. A medicine man rushed out of a nearby temple and claimed that the geologist had taken the chief's soul, and he would die. The quick-thinking geologist wadded the bouquet and made a pass toward the chief's mouth. He slapped him sharply on both cheeks and proclaimed that his spirit had returned. This did not passify the medicine man. He insisted the geologist was a devil-doctor who had brought an evil spirit to the village. Then he added a new twist: because of this evil, the chief's 16-year-old son, who was lying in the temple ill, would never get well. By this time, the situation had gotten out of hand, and the geologists realized they were in serious danger. They had to do something, so they asked to see the boy. Inside the temple, the boy was lying on a mat. He had a high fever, and his back was covered with huge infected blisters. Outside, the now apprehensive headhunters waited, apparently eager to do away with the geologists if the boy died.

The magician-geologist examined the boy's wound and noticed something strange. He gave the boy a local anesthetic with a hypodermic needle, then probed the wound with a sharp knife. Deep inside, he found an ugly thorn which had clearly been cut and deliberately placed inside the wound. The culprit could only be the medicine man.

He cleaned the wound, applied compresses soaked in warm water and antiseptic and waited. The three men took shifts to watch over the boy to make sure the medicine man could not undo their work. By the end of the week, the boy was entirely well; and the geologist and the chief were friends.

Undaunted by the close call, the geologists moved on to the island of Sumatra, traveling into the bush with a party of Malays led by a half-Chinese foreman. As they explored one area, they became aware of a pair of huge orangutans—wildmen-of-the-jungle, as the Malays called them. Orangutan skins and skeletons were worth a great deal of money at that time, and the foreman knew a collector in Java who paid well for such treasure. When the foreman begged for a rifle, the geologists refused and hid their guns. But the foreman discovered the hiding place and proceeded to shoot the female orang.

The enraged male apparently dropped out of a tree on top of him. What the others found were simply shreds. The rifle barrel was bent as if it were plastic; the stock was shattered. The foreman was literally torn limb from limb.

A trail of blood indicated that the male orang had taken his wounded mate deep into the jungle. Three nights later, the camp was awakened by unearthly screams. The male orang had returned. He kept out of sight, high in the trees, screaming all night. At dawn, the sounds stopped. The camp members were edgy, but they proceeded with their work. Then he struck. Three Malays who were hacking a clearing not far from the tents were attacked. The orang grabbed one, broke his neck and back and hurled him against a tree trunk. The others fled.

Night after night, the orang returned to berate the camp. Twice more he killed camp natives as they worked in the jungle. Finally, after ten days, the booming call grew more distant. The orang disappeared. One morning after the camp had settled down, one of the geologists and his assistant started up the trail to an oil seep. They were scanning the ground carefully when an uneasy feeling crept over them. They glanced up, both at the same instant. Not 20 feet away, the orang was moving along a thick branch in order to position himself to leap onto them. They reached for their automatics. The orang let out a scream, took two or three quick steps along the branch and hurled himself directly at the geologist. The geologist never remembered how he managed to get his automatic out of its holster, or how he managed to aim at the form. But he leaped backwards, firing twice. As the orang fell, its hands tore the shirt off the geologist's chest. The orang collapsed, dead, at his feet.

Tiger in the Eye

Late in 1938, Walter E. Nygren and I.K. Nichols, employees of the NNPM (N.V. Nederlandsche Pacific Petroleum Mig.), a subsidiary of the Standard Oil Company of California and The Texas Company (Caltex), were assigned to do surface geological work in an area northwest of Pakan Baru [Pekanbaru] in Central Sumatra.

"We established a camp in an area where there was a grove of durian trees," Nichols wrote. "These trees have a large, thorny fruit which has a very strong, pungent aroma and is a favorite food of the monkeys. We called it Camp Duri.

"Our house was constructed of poles with thatched palm sides and roof. The floor was made of split bamboo and was about three feet off the ground in order to avoid the many snakes, scorpions, etc. We slept on army cots with mosquito nets draped over them, to keep out the many insect pests.

Activity was picking up in Texas when H.H. Bradfield posed at Big Bend National Park, about 1937. Bradfield and the photographer, R.S. Powell, were both with The Texas Company at the time. R.C. Walther.

"The company provided each geologist with a Malay or Javanese servant boy who did the cooking, serving, washing and general housework. We were also provided with the very best food that good American dollars could buy. The supplies were brought in by boat from Pakan Baru to the village of Balaipungut, on the Mondau River, where we had a warehouse. From there everything was carried 'piggy-back' by coolies over elephant trails to our camp at Duri, a distance of about thirty kilometers. We had everything from hors-d'oeuvres to after dinner mints. The company followed this policy in order to protect and preserve the health and morale of the geologists, who had a mighty tough job under the best of conditions.

"The company had devised a method of working out the surface geology in central Sumatra which is peculiarly adapted to that area...It was found that good samples could be obtained by drilling deep hand-augur holes, and the approximate dips and strikes could be determined by making oriented punch cores in the same holes from which the hand samples were obtained."

Nygren cut straight, parallel trails, six kilometers apart, through the impenetrable jungles at approximately right angles to the trend of the regional folding. Along these trails at 50- to 200-meter intervals, auger holes were dug by hand. "It was usually necessary to drill down about five meters to penetrate the weathered zone and get into the fresh, unweathered strata," Nygren noted. "The holes were usually large enough for a man to climb down into via a rope and make readings on the bottom with a flashlight and Brunton compass."

If they were lucky, they could get oriented cores of bedrock. Nygren's crew dug 3,000 holes. (A similar technique had been used by Vincent Charles Illing on the island of Trinidad, when he evaluated the Naparima region. A meticulous researcher, Illing dug 9,000 pits, trenches and deep augur holes to draw up samples.)

"All of the auger holes, dips and strikes, and other pertinent data were plotted on a large scale map," Nichols wrote. "It was then that the geologic picture of the area would begin to take form, in the same manner that the picture on a blank kodak film takes form when immersed into the developing solution.

"Our days in the jungles began at daybreak, when we were awakened by the incessant howling of the wah wahs (Malay for black gibbons) as they began their search for a bite of breakfast to 'take the wrinkles out of their bellies.' Their reverberating howls would cascade down upon our palm house from every corner of the jungle, and further sleeping was impossible. I often profanely concluded that there were enough monkeys in one square mile of this jungle to fill all of the zoos from Hoboken to Hoquiam.

"After my morning coffee, I would have the head mandoer (foreman) take the coolie roll call, and then I would start out on the day's work at about 5:30 a.m. The morning hours are the best for working in the tropics, as the afternoons are usually terrifically hot".

Nichols had had no opportunity to learn to speak Malay before he was put in charge of

Laurence Brundall takes a break while core-drilling near Conroe, Texas, November 1937. L. Brundall.

Geologists who attended the Trinidad Geological Conference in April 1939 took along their summer whites, the better to enjoy the tropical weather. AAPG.

the drilling party. "During the first few weeks I underwent many exasperating experiences. I wasn't able to give orders to the coolies in Malay and they couldn't understand a word of English. Consequently, I had to rely upon sign language and a few key Malay words, which I had looked up in my pocket dictionary. I carried the dictionary with me every day and studied it diligently at night, so that in a few weeks I had learned enough Malay to supervise the work.

"Most of the coolies had never seen a pipe wrench or a pipe before and it took the 'none-too-bright' coolies a few days to learn how to couple and uncouple the ¾ inch pipe with a pipe wrench. Invariably they would turn the pipe the wrong way or cross thread it when coupling the pipe together. I had to show them how to do even the simplest task. I took the brightest of the coolies and made mandoers of them and placed one mandoer in charge of each drilling crew of four men. Only a few of the coolies could write, and I taught these men to label the sample bags and paint the number of the auger hole on a tree beside the hole. I gave each man a simple, routine job to do and after a few days the work went along smoothly.

"This area was considered to be bad tiger country and every day we saw their large tracks in the fresh mud along the trails. Some of these tracks were as large as a salad plate, and I often wondered what I would do if I met one face to face. One evening while coming in from work with the coolie along one of the jungle trails, we did meet a large tiger. We all stood in our tracks as if hypnotized. We stood thus for what seemed to me fully a minute. I had no gun but carried only a small Bowie knife. Each coolie carried a long wicked knife called a parang. One may shoot tigers from the height of a platform in a tree or an elephant's back with comparative safety; but when you come down to the ground the feeling comes home to you of the marvelous strength and activity that are combined in that beautiful frame. You know that when the occasion comes, the beautiful brute can come with lightning speed through the thick tangled jungle. Finally you realize that at close quarters a man is as helpless as a child against the over-powering weight and strength of an animal that can kill an ox with one blow. When you are on the ground looking a tiger in the eye, you realize all this with great vividness. The tiger merely looked us over, yawned, and finally strolled off nonchalantly into the jungle. Needless to say, we all heaved a big sigh of relief."

Water was another major obstacle in the jungle. The geologists walked, waded and paddled along thousands of miles of Indonesian rivers, searching for outcrops. Much of the work involved wading in water as much as four feet deep. Flat sticks were slapped sharply on the water surface to frighten away crocodiles. The early geologists spent months, day after day, with wet feet. In the evenings when they finally reached their tent-camp, the men would dry their bare feet over kerosene lamps and paint their toes and soles with potassium permanganate to keep the skin from peeling off. Some tried rubber hip-boots but they were generally too cumbersome.

Al Woodward (left) poses with Jim Brazil, an unidentified employee and Grady Davis at Mene Grande Company headquarters in Ciudad Bolivar, Venezuela, in 1938.
A.F. Woodward.

The office staff for Standard Oil Company of Egypt pose in Cairo in 1939 or 1940. Seated left to right: Carroll Cook, Cliff Armstrong, Dr. Bob Watson and J.P. "Jack" Gallagher. W.E. Wallis.

After many months of work, Nygren's crew discovered and mapped a large anticline about 15 kilometers long and five kilometers wide. It appeared extremely promising. Because the jungle soil was so thick, some of Nygren's auger-holes did not penetrate deeply enough to yield conclusive results in bedrock. To check his hypotheses, the company decided to drill a row of deeper holes across the postulated anticline with a hand-counterflush rig which could penetrate to 1,500 feet. Richard H. Hopper was sent to the area in 1939 as geologist on the job.

The hand-counterflush rig had a 30-foot steel tripod derrick which came in sections and could be easily assembled and disassembled as needed. Each section could be carried by two

Lunch "a la Richards"—a geologist tries a healthy repast on a Kansas Geological Society field trip to the Hartville Uplift, eastern Wyoming, in 1940. V.B. Cole.

men. It took eight men to man the pump, and it had to keep going 24 hours a day. If the drilling and water circulation were stopped, the sides of the hole were likely to cave in and stick the pipe. That meant work had to be done in three eight-hour shifts, the entire operation powered exclusively by human muscle.

"It amazes most American oil people that we could drill and core to 1,500 feet without any power except men's bodies," Hopper wrote.

Hopper was also stationed at Pekanbaru. The counterflush equipment and supplies were moved by barge up the Siak River from Pekanbaru to Palas, and then carried in to the northeast, along one of Nygren's trails to the four drill sites that had been chosen. The food was carried the same way, and it required a two-day trip by loaded porters from Palas.

It took Hopper's crew seven months to drill the four counterflush holes. In 1940, a seismic crew was moved in to the area to check the structure further. The seismic crew used the trails cut by Nygren to get into the area.

When that crew's report also was favorable, the company had to face the transportation problem. In order to move in a drilling rig capable of penetrating about 4,000 feet, they would have to build an access road for trucks. The road was 35 kilometers over hilly terrain. It was all done by hand labor, and was not completed until September 1941.

The first barge-load of trucks and drilling equipment arrived shortly after that. The drillers put up a camp of tents and thatched huts at the drilling site, and began assembling parts of the rig.

But before they could commence drilling, the Japanese attacked Pearl Harbor. Hopper happened to be in Timor doing geological work the day Pearl Harbor was bombed. "One of the objectives of the Japanese was to obtain control of the resources of Southeast Asia," Hopper wrote. "In particular, they wanted to take the oil fields and refineries in an undamaged state." To accomplish this, a Japanese force which had embarked from Saigon was parachuted into the Palembang refineries in February 1942, just as Singapore was falling to another Japanese force which had come down overland on the Malay Peninsula. Japanese troops moved almost immediately into Malaya, the Philippines and the Indonesian islands. Caltex drillers and geologists were instructed to abandon the Minas wellsite as well as the Duri and Sebanga oil fields, which were not yet on production. As the Japanese invaded Java and Sumatra, Caltex expatriates were either evacuated or joined the armed forces of their own countries. The Indonesians returned to homes and families and awaited the arrival of Japanese forces.

Within a month, Hopper went to Jakarta and joined the U.S. Army, serving in South Sumatra, Java, Australia, Irian (New Guinea) and the Philippines. Immediately after the Netherlands East Indies surrendered to Japan on March 9, 1942, the Japanese began consolidating their hold on the country. They needed aviation gasoline, motor gasoline, diesel fuel and other products to support their war effort. They set up a major general, with headquarters in Palembang, to administer the entire petroleum industry of Sumatra. A division of oil technicians was at the disposal of the major general.

Not all of South America was jungle. In 1939–1940, Victor Oppenheim crossed the Nevada de Cocuy, Colombia (elevation 18,000 feet). The natives who accompanied him usually chewed coca leaves to make breathing and climbing easier. V. Oppenheim.

Sherman A. Wengerd runs an alidade while plane tabling on Ordovician rocks, Ramshorn Mountain, Salmon River Range, Idaho, July 1938. Wengerd would later serve as president of AAPG, in 1971.
S.A. Wengerd.

When W.S. Pike, Jr. (left) and Larry Brundall (right) worked for Shell Oil Company in Many, Louisiana, in 1939, they occupied close quarters.
L. Brundall.

division of oil technicians was at the disposal of the major general.

They drilled a few new wells in established oil fields, but only one wildcat well—Minas No. 1—using the rig and equipment that N.P.P.M. had already set up at the site. It was the only wildcat well ever drilled by an army of occupation. In December 1944, it was completed at a depth of 2,017 feet. "What a discovery!" Hopper wrote. It proved to be the second largest oil field in Sumatra, with several billion barrels of reserves. But the outside world did not hear about it until after the war.

Mountain High

To the geologist, there was nothing more beautiful or more sacred than the land. The young woman who married such a man had to be either understanding and tolerant or share the same feelings. One couple who understood and shared that love for the earth was Chester K. Wentworth and his bride-to-be Mildred Porter. They decided there was no more appropriate place to have a wedding ceremony performed than in the outdoors they loved. In August 1920, they found an accommodating minister and climbed high into the mountains of Virginia for their special wedding. The bride and groom were both dressed in appropriate hiking attire.

Amid Falling Bricks

In 1925, Bailey Willis of Stanford University, felt sure that a major earthquake was imminent in California's Santa Barbara area. Most ignored him when he suggested that such an event would occur within the next six months. But the earthquake occurred right on schedule; and although Willis denied the story, it was said that he was there taking down notes for posterity while the bricks were still falling.

A Matter of Luck

When Fred Sutton wrote his wife, Anne, from South America in 1927, he noted the universal plight of the geologist and how outside forces could shape his career. "It's entirely a matter of luck where one is sent to work, & if there is anything there he finds it & gets the credit. If there is nothing there, he condemns it, and the report is just as valuable as the positive one although the general public feels that the one who makes a location and gets oil is the better geologist. In some cases the geologist himself gets to feeling that way."

Turning Something Up

When E.L. Estabrook wrote about conditions in the Rocky Mountain region in 1925, the principles he saw at work were true for other areas as well. "Progress in prospecting...sometimes seems slow and the results all too meager in proportion to the money that has been spent," he wrote, "but wildcatting is likely to continue for many years because something is always turning up to sustain the interest."

Spirit of Adventure

When Victor Oppenheim explored Porongal, a wild area of Peru, in the 1930s, he took with him seven men and 10 riding and packing animals. Only one of the men had ever been into Porongal. Some of the men were quite young, others elderly; but they were all eager to find out what was at the foot of the mountains. "It was remarkable how these men were ready to embark on this new venture, leave their wives and mountain homes and follow a stranger, as I was to them, into a wild jungle country," Oppenheim wrote in *Exploration East of the High Andes*. "It was a place which they had heard of since childhood as a bad and fearsome country, where men had gone and never returned. Yet, enthused more by adventure than by the modest pay, these men were willing and eager to face whatever might come."

Oppenheim was convinced that this behavior was a part of man's basic nature. "I believe this spirit of adventure is deeply rooted in man, and is almost atavistic in many of us today, as it was millenia ago. This is the same spirit that made the crusaders follow the banners, or the thousands of conquistadores converge on the New World or later the frontiersmen to cross the great plains of North America. This is the spirit that moves every daring enterprise, every step off the beaten track and into the unknown."

The Important Thing

When Victor Oppenheim was asked to join Yacimientos Petroliferous Fiscales (YPF), the Argentine government petroleum corporation, he was already chief engineer and geologist of an expanding mining company. His current position afforded him all the comforts of big-city life in Buenos Aires, pleasant and well regulated. What YPF offered was the position of field geologist, to lead a geological expedition to the upper Bermejo River.

He resigned his post and struck out with YPF. "I believe my choice was correct," he wrote in *Exploration East of the High Andes*. "I exchanged the superficial pleasures and commodities of a profitable position and city life, for the hard and rude life of geological field exploration in regions of inaccessible mountains and tropical jungles. To most modern young men this

Early Schlumberger logging equipment was still in use in eastern Venezuela in 1939. Logging core holes for Socony Vacuum (Mobil) are Charles Rosoff, Schlumberger engineer, and Myron Gerson, driller. "The log was in two parts, pencil drawn on paper; an SP and a Resistivity log," wrote Robert A. Bishop. "The operators watched galvanometers, keeping the needle on the midpoint. If the needle moved to the right, the operator cranked the wheel counterclockwise to bring the needle back to the midpoint, and vice versa. A pencil attached to an eccentric on the wheel drew the curve as the paper moved past. The completed log looked similar to log curves in use today and worked well as a correlation tool. The Schlumberger operator logged the Resisitivity curve, which was usually the most erratic, and the geologist operated the SP curve." R.A. Bishop.

A well blow-out—rig with stand of drill pipe falling over and plume of drilling mud and/or oil and gas rising vertically from borehole; Louisiana.

would have seemed like a bad bargain. Yet at the time, there was nothing more important to me than the exploration of that little-known region of the Argentine Andes. Money, comfort and security were inconsequential. The quest of knowledge, the geology of that particular area, its composition and structure—this became to me most important.

"...So it was that I set out on another trip."

Too Long Too Deep

In 1934, when C. Hewitt Dix began making field trips for Humble Oil Company, much of the time was spent burying large geophones a meter deep in hard west Texas crust. Dix was with the crew off and on for more than two months. The work and the methods were so time-consuming that in all that time, they never reached the edge of the reservoir they were trying to map.

Fresh Enough for Scrambled

In 1936, W.E. "Bill" Wrather was in California working on the famous Kettleman North Dome Association law suit, with Robert Moran, Johnny Galloway, Ralph Reed and Martin van Couvering. Moran called the affair "the Geologist's W.P.A."

"Bob and I took advantage of the opportunity to make weekend trips together—to Death Valley; to mineral-bearing localities within reach," wrote Wrather.

William R. Moran, Robert's son, was also included on some of the trips. "One trip we had spent the night in sleeping bags and decided to save time by getting breakfast at a small desert cafe we had spotted, rather than getting out the cooking gear," Moran wrote. "Bill ordered his eggs scrambled, while I followed my dad's lead, and we asked for ours poached. Bill's order came through promptly, and we told him not to wait for us, so he put it all away without further comment. My dad and I waited...and waited...and waited, and we began to get smart remarks from Wrather about his pull with the waitress, undoubtedly caused by his good looks and gentlemanly manners. Finally, a distraught waitress appeared. 'Would you gentlemen just as soon have your eggs cooked in some other way?' she asked. 'The cook has just tried and tried, but he can't find any eggs fresh enough to poach.' We decided to pass on the eggs, and Bill told me years later that was the last time he had eaten scrambled eggs!"

John Marshall (left) and Burton E. Ashley stroll downtown in Wichita, Kansas in 1940. It was a far cry from Ashley's African adventures. B.E. Ashley.

J.E. Pemberton(?) found the airplane a good way to get around and a better way to explore the geology of the land. American Heritage Center, University of Wyoming.

Blowouts were often spectacular and could be seen for miles. Field geologists would often travel an hour or more to see what had happened. This blowout was south of Taft, California, April 1939. E.W. Owen/Mirva Owen.

Paleontology students examine an Eocene laurel log at Iowa Flats, northern Sierra Nevada, California, in 1939. Left to right: Frank Peabody, Allan Bennison, Dorothy Schoof and Bradford Orr.

Oklahoma City Geological Society members piled into their cars, for a field trip to Oklahoma Panhandle and New Mexico, in May 1941. V.B. Cole.

End of an Era

At the end of the 1930s, as war began to rumble through the world, the field of petroleum geology was still in its adolescence. Technology had expanded from hammer compasses and sturdy ruled walking-sticks—recommended by one English geologist in the 19th Century—to seismology and well logging. Mules and horses had been replaced by automobiles and airplanes except in most remote corners of the earth.

Most of the petroleum geologists at the beginning of World War II knew their compatriots. They had run into each other in remote deserts and jungle camps; they had read of each other's exploits in company reports. They were friends, or friends of friends, developing a camaraderie despite the distances and perhaps years that passed between meetings. There was a fellowship of purpose even in company rivalries.

Times had often been difficult during the 1930s as the cyclical nature of the industry made itself felt. But the war was changing attitudes. The expertise of petroleum geologists was needed to help with reconnaissance, mapping, intelligence work and other strategic needs.

The petroleum geologists joined the war effort, eager to be of service. Hundreds of young men who strode off to war did so with a double eye—one aimed for danger and another at the earth itself. Perhaps they would someday return in some future search for energy.

They did not know then that their efforts would completely change the world and alter the petroleum industry.

Select Bibliography

Amerada Hess, letters. Donated to AAPG.

American Association of Petroleum Geologists Bulletin. From inception to 1984.

Bailey, S.W. *The History of Geology and Geophysics at the University of Wisconsin-Madison, 1848-1980.* Madison, Wisconsin, University of Wisconsin-Madison, Department of Geology and Geophysics, 1981.

Baldwin, Tom. *40 Years: The Education of a Geologist.* Los Angeles, L.A. Basin Geological Society, Pacific Section AAPG, 1952.

Baldwin, Tom. "*Tales from the Rig Floor.*" Unpublished Ms.

Beeby, Arthur Thompson. *Black Gold.* Garden City, New York, Doubleday, 1961.

Crump, Irving. *Our Oil Hunters.* New York, Dodd, Mead and Co., 1948.

Eby, J. Brian. *My Two Roads.* Houston, Gulf Publishing Company, 1975.

Ellis, William Donohue. *On the Oil Lands with Cities Service.* Cities Service Oil and Gas Corporation, 1983.

Facts and Fantasies of the Oil Patch. Desk and Derrick of Oklahoma City, 1975.

Fanning, Leonard M. *The Rise of American Oil.* New York, Harper & Bros, 1936.

Gould, Charles. *Covered Wagon Geologist.* Norman, University of Oklahoma Press, 1959.

Ham, Elizabeth A. *A History of the Oklahoma Geological Survey, 1908-1983.* Oklahoma City, Oklahoma Geological Survey, 1984. [OGS 83-2.]

Hamilton, Charles Walter. *Early Day Oil Tales of Mexico.* Houston, Gulf Publishing Co., 1966.

Hares, Charles J. "*History of Wyoming Oil,*" unpublished manuscript, 1956. 35 p. (University of Wyoming).

Hopper, Richard H. "The Discovery of Indonesia's Minas Oilfield." *Oil—Lifestream of Progress (1),* Caltex Petroleum Corporation, 1976.

Hopper, Richard H. "Fifty Years of Exploring for Oil in Indonesia." *Oil—Lifestream of Progress (3),* Caltex Petroleum Corporation, 1974.

Hopper, Richard H. "Petroleum in Indonesia: History, Geology and Economic Significance." Lecture, Indonesian Council of the Asia Society, December 15, 1971.

The Humble Way. May-June 1957. Vol. 13, No. 1.

Knowles, Ruth Sheldon. *The Greatest Gamblers*. Norman, University of Oklahoma Press, 1978.

Magnolia Oil News. Magnolia Oil Company. April 1931.

McCrary, E.W. *Reminiscences*. Unpublished notes, 1964, E.W. Owen Collection, University of Wyoming.

McIntyre, James. *Oil & Gas Journal*, Nov. 11, 1926, p. 32.

Morley, Harold T. *A History of the American Association of Petroleum Geologists: The First Fifty Years*. Tulsa, AAPG, 1966.

Oppenheim, Victor. *Exploration East of the High Andes: From Patagonia to the Amazon*. New York, Pageant Press, 1958.

Owen, Ed. *Trek of the Oil Finders*. Tulsa, Oklahoma, American Association of Petroleum Geologists, 1975. (Memoir 6)

Pacific Oil World. August 1983, Vol. 75, No. 8, p. 8-26.

Petroleum Production Pioneers Oral History Project. Transcription of material which belongs to Joe J. Oliphant, Bakersfield, California. Provided by W.R. Moran of Molycorp, Inc.

Pratt, Wallace E. *Oil in the Earth*. Lawrence, University of Kansas Press, 1942.

Reinholt, Oscar H. *Oildom: Its Treasures and Tragedies*. Philadelphia, David McKay, 1924.

Scott, Otto J. *The Exception: The Story of Ashland Oil & Refining Co.* New York, McGraw-Hill.

Spence, Hartzell. *Portrait in Oil: How the Ohio Oil Company Grew to Become Marathon*. New York, McGraw-Hill Book Co., 1967.

Tait, Samuel W., Jr. *Wildcatters: An Informal History of Oil-Hunting in America*. Princeton University Press, 1946.

Tinkle, Lon. *Mr. De: A Biography of Everette Lee DeGolyer*. Boston, Little, Brown Co., 1970.

University of Wyoming, Petroleum History Collection. Extensive collection, including letters, diaries, notes and unpublished manuscripts.

Weeks, Lewis G. "....a lifelong love affair." Westport, Connecticut, Anne Sutton Weeks, 1978.

Woolnough, W.G. "*The Parliament of the Commonwealth of Australia, 1929-1930-1931. Report on Tour of Inspection of the Oil-Fields of the United States of America and Argentine, and on Oil Prospects in Australia*." nd, location, etc.

Youngquist, Walter. *Over the Hill and Down the Creek*. Caldwell, Idaho, Caxton Printers, 1966.

Acknowledgments

The following people and institutions generously contributed to this book with their time and energy, and often with the loan of photos, documents, and other materials. Many items were sent from personal collections, while others came from museums, libraries, and industry archives.

We gratefully acknowledge the following contributors for their invaluable help in preserving this segment of petroleum geology's rich and colorful past.

American Association of
Petroleum Geologists (A.A.P.G.)
Tulsa, Oklahoma

American Museum
of Natural History
New York, New York

Aminoil
Houston, Texas

Burton E. Ashley
Washington, D.C.

Floyd Ayers
San Antonio, Texas

A. Bennison
Tulsa, Oklahoma

R. A. Bishop
Denver, Colorado

The Blakey Group
Tulsa, Oklahoma

M.N. Bramlette (deceased)

George Hansen Collection,
Brigham Young University
Provo, Utah

Laurence Brundall
Santa Barbara, California

W. D. Chawner
Del Mar, California

Cities Service Company
Tulsa, Oklahoma

L. M. Clark/
A.A.P.G.

Joe M. Clark
Fayetteville, Arkansas

Willard J. Classen
Menlo Park, California

Virgil B. Cole (deceased)

Robert H. Dott, Sr. (deceased)

J.B. Eby
Gulf Publishing

J.E. "Brick" Elliot (deceased)

Mrs. Rolf Engleman
(the late Rolf Engleman's photos)
Oklahoma City, Oklahoma

C. A. Evans
Aminoil

N. L. Falcon
Chiddingford, Surrey, England

Michel T. Halbouty
Houston, Texas

K. C. Heald
U.S.G.S.

Bela Hubbard (deceased)

Humble Way
Houston, Texas

Robert E. King (deceased)

H.M. Kirk (deceased)

H.R. Kramer

T.A. Link (deceased)

M.D. Maravich
Arcadia, Oklahoma

Oklahoma Geological Survey
Norman, Oklahoma

Victor Oppenheim
Dallas, Texas

Mirva C. Owen
(the late Ed Owen's photos)
San Antonio, Texas

Phillips Petroleum
Bartlesville, Oklahoma

Sidney Powers/
American Heritage Center,
University of Wyoming

W. Armstrong Price
Corpus Christi, Texas

E. M. Spieker
U.S.G.S.

University of Missouri at Columbia
(lent by Tom Freeman)
Columbia, Missouri

William E. Wallis
Santa Barbara, California

Roy C. Walther
New Orleans, Louisiana

Mrs. L. G. Weeks
(the late Lewis G. Weeks's Memoirs)
Westport, Connecticut

Sherman A. Wengerd
Albuquerque, New Mexico

K. D. White/
American Heritage Center,
University of Wyoming

Claude Williams
Amerada Petroleum Company

A. F. Woodward
Tigard, Oregon